Lecture Notes in Artificial Intelligence 518

Subseries of Lecture Notes in Computer Science
Edited by J. Siekmann

Lecture Notes in Computer Science
Edited by G. Goos and J. Hartmanis

J. G. Williams

Instantiation Theory

On the Foundations of Automated Deduction

Springer-Verlag
Berlin Heidelberg New York
London Paris Tokyo
Hong Kong Barcelona

Series Editor

Jörg Siekmann
Institut für Informatik, Universität Kaiserslautern
Postfach 3049, W-6750 Kaiserslautern, FRG

Author

James G. Williams
The MITRE Corporation, M/S A129
Burlington Road, Bedford, MA 01730-0208, USA

CR Subject Classification (1991): I.2.3

ISBN 3-540-54333-3 Springer-Verlag Berlin Heidelberg New York
ISBN 0-387-54333-3 Springer-Verlag New York Berlin Heidelberg

Printed in Germany

Typesetting: Camera ready by author
Printing and binding: Druckhaus Beltz, Hemsbach/Bergstr.
2145/3140-543210 - Printed on acid-free paper

PREFACE

Instantiation theory is the study of instantiation in an abstract context that is applicable to most commonly studied logical formalisms. This book begins with a survey of general approaches to the study of instantiation, as found in tree systems, order-sorted algebras, algebraic theories, composita, and *instantiation systems*.

A classification of instantiation systems is given, based on properties of substitutions, degree of type strictness, and well-foundedness of terms. Equational theories and the use of typed variables are studied in terms of quotient homomorphisms and embeddings, respectively. Every instantiation system is a quotient system of a subsystem of first-order term instantiation.

A general unification algorithm is developed as an application of the basic theory. Its soundness is rigorously proved, and its completeness and efficiency are verified for certain classes of instantiation systems. Appropriate applications of the algorithm include unification of first-order terms, order-sorted terms, and first-order formulas modulo α-conversion, as well as equational unification using simple congruences.

I am indebted to William Farmer for acquainting me with the literature on unification algorithms, for help in formulating the basic theory, and for valuable advice regarding its development. I also wish to thank Hans-Jürgen Bürckert, Dale Johnson, John Stell, and an anonymous referee for valuable feedback on its presentation. This work was sponsored by the Rome Laboratories, Griffiss Air Force Base, Rome, NY 13441, under the direction of John C. Faust, COAC.

Bedford, MA
June 1991

James G. Williams

TABLE OF CONTENTS

SECTION 1
INTRODUCTION

Instantiation theory axiomatizes properties of instantiation found in most formal reasoning systems. Systems covered include full predicate calculus, order-sorted generalizations of open predicate calculus, lambda calculi, and the "rational term" systems used in some Prolog implementations, as well as quotients of these systems induced by equational identities.

This work focuses on structural, or algebraic properties of instantiation. It seeks to achieve breadth of applicability by avoiding a priori constraints imposed by syntax and semantics and to achieve elegance by avoiding inessential details through the use of abstraction.

This paper is written for a general mathematical audience and is relatively self-contained. Its purpose is to contribute to a broad but useful foundation for a general theory of deduction. As a test of utility, instantiation theory is applied to the construction of a general unification algorithm.

MOTIVATION

The present theory of instantiation systems stems indirectly from a consideration of design requirements for software reasoning systems [WILL87]. A good deal of sophistication seems necessary in order to meet such requirements [cf FARM90, FARM90b], and it is known that first-order logic does not adequately support the necessary semantic conventions [WILL90].

Although a great many logical formalisms have now been identified, only a few of them have been widely studied, and there has as yet been no general theory of logics that can facilitate the development of new formalisms or the comparison of existing formalisms.

Much of the work in automated deduction has been confined to the study of first-order terms and quantifier-free formulas, and much of this work can be understood in terms of instantiation. These observations suggest a general theory of instantiation as the starting point for determining the general applicability of what is already known about deduction in particular formal systems.

ORGANIZATION AND CONTENT

Section 2 reviews basic properties of monoid actions [CLIF67], gives relevant examples based on instantiation, and presents the unification problem in its simplest, most general form. Section 3 introduces *instantiation systems* and compares them with other general approaches to the study of instantiation. Instantiation systems allow for (but do not require) typed variables, the identification of terms under congruence relations, and the avoidance of "occurs checks" through the use of rational trees.

Section 4 classifies instantiation systems according to various properties and sets forth the principal advantages of each classification property. Section 5 studies subsystems, quotient systems and, more generally, homomorphisms between systems. Its characterization of congruence relations is the basis for most of the examples presented in this paper. Section 6 augments instantiation systems with a set of *basic syntactic constructs*, investigates the construction of terms from variables and constructs, and gives results on finite tree-implementations. These show, in particular, that every countable instantiation system is a quotient system of a subsystem of the term-language of predicate calculus.

Section 7 presents a unification algorithm and proves its soundness. Sections 8 and 9 investigate issues relating to completeness and computational complexity, respectively. Finally, Section 10 discusses the future development of the theory. Appendix A presents the results of the main implementation and optimization strategies given in Sections 8 and 9.

In the remaining sections, explanatory remarks are separated from surrounding mathematical presentation by a horizontal line. Initial and defining occurrences of new terms are set in italics for later reference.

RELATED WORK

Several previous works deal with general theories of instantiation. Nerode's theory of "composita" contributes several basic ideas to the present development, despite the fact that his work predates unification theory and is restricted to "untyped" systems [NERO59]. The same is true of Elgot's work on algebraic theories [ELGO75]. Category-theoretic formalizations of untyped instantiation have also been given by Rydeheard, Burstall, and Stell [RYDE86, RYDE87]. All of these treatments are sufficiently general as to encompass the tree instantiation systems of Courcelle, as well as quotients of first-order term instantiation.

Steinbrüggen used a theory of monoid actions similar to that given in Section 2 as the starting point for an abstract theory of program transformations [STEIN81], and Goguen has given a generalization of algebraic theories that allows for typed variables [GOGU88]. These more general treatments allow for the instantiation concepts found in order-sorted algebras [GOGU87]. Finally, instantiation systems are equivalent to the independently discovered "unification algebras" of Schmidt-Schauss and Siekmann [SCHM88].

The idea of using quotient systems to study equational unification appears already in the works of Pietrzykowski and Jensen [PIET72, JENS73]. The unification algorithm developed in Sections 7 — 9 allows a systematic comparison of this idea with the more pedestrian approach of working directly on concrete syntactic objects. Other ideas that have been incorporated into this development include the following:

triangular form	[MART82, MART86, SCHM88]
systems of multiequations	[MART82, JAFF84, MART86, SCHM88]
reliance on construct unifiers	[FRAN88, STICK81]
rigid-term optimization	[HUET75]
classification properties	[BURC87, SCHM88]
flexible typing	[MESE87, WALT86, SCHM86]
variable-renaming avoidance	[ROBI79, MART82, JAFF84, SCHM88].

The algorithm schema given in Section 7 is distinguished from these previous works by the fact that these ideas are now integrated into a single, unified development.

Although both the present work and that of Schmidt-Schauss and Siekmann investigate the same theory and address many of the same issues, there are major differences in purpose and content. Their earlier work is an initial investigation whose purpose is to test the utility of the theory for studying unification. The purpose of the present work, by way of contrast, is to provide a solid axiomatic foundation for the study of deduction. In this work, basic definitions and theorems are presented in their strongest useful form. Their sharpness is validated, in most cases, by accompanying counterexamples and exercises. The presented results include appropriate generalizations of relevant theorems from algebraic theories and similar previous studies, in order to provide an understanding of what is known or readily knowable.

Both investigations focus on the unification strategies of Martelli and Montanari [MART82]. The Schmidt-Schauss work provides an elegant, deterministic unification algorithm that is complete for a limited class of unification algebras, but does not address implementation issues or computational complexity. The present treatment covers a much broader class of instantiation systems and shows why some Martelli-Montanari strategies that were later rejected in the case of first-order term unification [JAFF84] are still needed. Finally, it includes a full, general investigation of the ingenious Martelli-Montanari implementation strategies and their computational complexity.

SECTION 2
BACKGROUND

Most instantiation mechanisms have the following properties: substitutions form a monoid under composition, and the application of substitutions to syntactic objects is a faithful, unital, right monoid action. The most common exception is that instantiation may be a partial operation. These observations form a natural starting point in the development of a theory of instantiation.

The following paragraphs review basic facts about monoid actions, introduce the unification problem in its simplest, most general form, and informally comment on its ramifications. Motivating examples are given at the end of this section. The last three of these are conceptually significant, because they give some hint as to the leeway which exists between instantiation and semantics.

MONOID ACTIONS

It is usually possible to think of substitutions as functions from O to O, by identifying each substitution σ with the function σ' given by $\sigma'(t) = t\sigma$. A potential problem with this representation is that, if the associated monoid action isn't *faithful*, we could identify two different substitutions with the same function. Another problem is that this representation falsifies the expected relationship between subsystems and embeddings. Although later sections use a different representation for faithful monoid actions, it will be useful to pursue this idea briefly, in order to study faithfulness and to review standard facts about monoid actions.

Any monoid action has the same terms and *instance relation* as a canonically induced faithful *quotient* action, and *unital* monoid actions have a simpler basic theory. For these reasons, we study only unital, faithful monoid actions in later sections. Faithfulness is equivalent to the existence of a *discriminating* set V. This fact allows us to assume a particular structure for substitutions that is well suited to the study of subsystems in Section 5. Thus, as a matter of convenience, we concentrate on what may be called actions *by substitutions on* V.

A *right monoid action* [HOWI76] (also known as a right *operand* [CLIF67]) is an algebraic structure $\mathcal{A}$ with the following form which satisfies axioms (1) through (3) below.† If $\mathcal{A}$ also satisfies axioms (4) or (5), then $\mathcal{A}$ is said to be *unital* or *faithful*, respectively.

Structure

- O : a set whose elements are referred to as *terms* (or *syntactic objects*).
- S : a set whose elements are referred to as *substitutions*.
- ϵ : an element of S, called the *null* substitution.
- $\bullet$: a function from S × S to S, called *monoid composition*.
- $\ast$: a function from O × S to O, called *instantiation*.

Notation

$s, t, s_1, t_1, \ldots \in O$.
$\sigma, \tau, \sigma_1, \tau_1, \ldots \in S$.
We write $\sigma\tau$ rather than $\bullet(\sigma, \tau)$, and $t\sigma$ rather than $\ast(t, \sigma)$.

Axioms

1 (*monoid associativity*). $(\mu\sigma)\tau = \mu(\sigma\tau)$.
2 (*monoid identity*). $\sigma\epsilon = \epsilon\sigma = \sigma$.
3 (*action associativity*). $(t\sigma)\tau = t(\sigma\tau)$.
4 (*unital identity*). $t\epsilon = t$.
5 (*faithfulness*). $\sigma = \tau$, provided $t\sigma = t\tau$, for all t.

Define $s \leq t$ iff $s\sigma = t$, for some σ. If $s \leq t$, we also say that s *generalizes* t and that t is an *instance* of s. We refer to $(\leq)$ as the *instance* relation. It is a *preordering*: in other words, it is reflexive and transitive.

Given any right monoid action $\mathcal{A}$, we define a *representation by transformations*, $\mathcal{A}'$, to be the structure with the following components:

$O' = O$.
S' is the set of all functions σ', for $\sigma \in S$, where $\sigma'(t) = t\sigma$, for each term t.
ϵ' is the identity function on O.
$\bullet'$ (functional composition) is given by $\bullet'(\sigma', \tau')(t) = \tau'(\sigma'(t))$.
$\ast'$ (functional application) is given by $\ast'(t, \sigma') = \sigma'(t)$.

This is the representation of faithful monoid actions which was used in the development by Schmidt–Schauss and Siekmann [SCHM88].

For any right monoid action, $\mathcal{A}$, let $\approx$ be the equivalence relation on substitutions defined by $\sigma \approx \tau$ iff $\sigma' = \tau'$. (Clifford and Preston refer to this as the *kernel* congruence.) A *discriminating* set for a right monoid action is a subset V of O such that whenever $v\sigma = v\tau$ for all v in V, we have $\sigma = \tau$.

† Substitutions are written on the right, because of their similarity to parameter bindings. This similarity is particularly relevant to the implementation strategies given in Section 6.

Proposition 2.1. For any right monoid action $\mathcal{A}$, we have,

A. $(\ )'$: $\mathcal{A}$ ->> $\mathcal{A}'$ is a surjective algebraic homomorphism.

B. $\mathcal{A}'$ is isomorphic to the quotient system $\mathcal{A}/\approx$ constructed in the obvious way.

C. $\mathcal{A}'$ is always faithful.

D. The following are equivalent:

  i) $\mathcal{A}$ is faithful.

  ii) $\mathcal{A}$ has a discriminating set.

  iii) $\approx$ is the identity relation.

  iv) $\sigma\tau$ is always the unique γ such that for all t, $t\gamma = (t\sigma)\tau$.

E. $\mathcal{A}$ and $\mathcal{A}'$ have the same instance relation.

Proof. Assertions A, B, C, and parts of D may be found in [CLIF67]. To show, for example, that (iv) implies (i), suppose $t\sigma = t\tau$, for all t. By assumption, $\epsilon\sigma$ is the unique substitution, γ, such that $t\gamma = (t\epsilon)\sigma$, for all t. Both σ and τ satisfy this definition. Hence $\sigma = \tau$. Finally, Assertion E is trivially equivalent to the observation that $s\sigma = t$ iff $\sigma'(s) = t$. □

In the theory of unital, faithful monoid actions, axioms (1) and (2) are unnecessary. The monoid-associativity axiom, for example, is an immediate consequence of the faithfulness and action-associativity axioms. Moreover, • and ϵ are definable: condition (iv) of Proposition 2.1D defines •, and the unital-identity axiom defines ϵ. Examples of these systems are thus identifiable in terms of the reduced structure (O, S, $*$).

Let V be a discriminating set for a unital faithful action $\mathcal{A}$. Let $f(\sigma) = \sigma'|\{v \in V \mid v\sigma \neq v\}$. Since V is a discriminating set, f is one-to-one. Consequently, f induces an isomorphism from $\mathcal{A}$ to a monoid action with the following additional properties: Each substitution σ is a partial function from V to O; $\sigma(v) = v\sigma$, for each v in the domain of σ, and the domain of σ is the set of all v such that $v\sigma \neq v$. We refer to a unital, faithful action with these properties as an *action by substitutions on* V.

A *term congruence* is an equivalence relation E on O such that s E t implies $s\sigma$ E $t\sigma$, for all s, t, and σ. Each term congruence E induces a corresponding substitution equivalence $\approx$, defined by $\sigma \approx \tau$ iff $t\sigma$ E $t\tau$, for all terms t. We write O/E and S/$\approx$ for the set of all E-equivalence and $\approx$-equivalence classes of O and S, respectively.

Exercise 2.2. A term congruence E on a unital, faithful monoid action induces a quotient unital, faithful action with instantiation operation $*/(E \times \approx)$ from $O/E \times S/\approx$ to O/E.

THE UNIFICATION PROBLEM

We briefly consider *unification*, first for an arbitrary monoid action, and then for the case of *first-order term instantiation*. After this, we informally consider some ramifications of the *unification problem*, both in general and for the particular case of first-order terms.

For any given monoid action, one can ask whether two terms, t_1 and t_2, have a *common-substitution* instance, that is, whether there is a substitution σ such that $t_1\sigma = t_2\sigma$. In this case, σ is referred to as a *unifier* of t_1 and t_2. Clearly, if σ is a unifier of t_1 and t_2, then so is $\sigma\eta$, for any η. A *most general* unifier σ has the property that, for every unifier τ, and every relevant variable v, $v\sigma \leq v\tau$. (This definition is motivated by Exercise 3.12D and Example 5.3A.)

The objects, substitutions, and instantiation operation of *first-order term instantiation* are just those associated with the term language of predicate calculus [ANDR86, LOVE78, ENDE72]. These form a monoid action in which the set of variables is a minimal discriminating set. As a matter of notation, we shall take the variables of predicate calculus to be the boldfaced symbols $\mathbf{x}$, $\mathbf{y}$, $\mathbf{x}_1$, $\mathbf{y}_1$, ... We assume countably infinite sets of individual constants and function symbols of each arity (including binary operators such as **+** or **−**).

Exercise 2.3. For first-order terms, the problem of finding common instances is different from, but equivalent to, the problem of finding unifiers and common-substitution instances:

A. The terms $\mathbf{x}$ and $\mathbf{x + 1}$ have a common instance, namely $\mathbf{x + 1}$, but do not have a common-substitution instance.

B. If s and t have no common variables, then s unifies with t iff s and t have a common instance.

C. If v doesn't occur in s or t, then s unifies with t iff the terms s **+** t and v **+** v have a common instance.

D. If s and t have variables in common, one can still use a unification algorithm, together with an appropriate variable-renaming strategy, to find common instances.

Exercise 2.4. Regarding first-order term instantiation,

A. Two terms of the form

$$r = \mathbf{f}(\ldots \mathbf{f}(\mathbf{x}_1, \quad \mathbf{x}_2), \ldots, \mathbf{x}_n),$$
$$s = \mathbf{f}(\ldots \mathbf{f}(\mathbf{1}*\mathbf{1}, \mathbf{x}_1*\mathbf{x}_1), \ldots, \mathbf{x}_{n-1}*\mathbf{x}_{n-1}),$$

have a common-substitution instance

$$t = \mathbf{f}(\ldots \mathbf{f}(\mathbf{1}*\mathbf{1}, \mathbf{1}^4), \ldots \mathbf{1}^{2^n}),$$

given by a most general unifier σ^n, where

$$\sigma = \mathbf{1}{*}\mathbf{1}/\mathbf{x}_1 + \mathbf{x}_1{*}\mathbf{x}_1/\mathbf{x}_2 + \ldots + \mathbf{x}_{n-1}{*}\mathbf{x}_{n-1}/\mathbf{x}_n$$

is the substitution that maps $\mathbf{x}_1$ to $\mathbf{1}{*}\mathbf{1}$, $\mathbf{x}_2$ to $\mathbf{x}_1{*}\mathbf{x}_1$, and so forth.

B. The above σ^n may be characterized as the unique substitution σ^* such that $\mathrm{dom}(\sigma^*) \subseteq \mathrm{dom}(\sigma)$ and $\sigma\sigma^* = \sigma^*$.

The *unification problem* for a given monoid action is to find an algorithm that, given two terms as input, will decide if they have a unifier and, if so, return a most general unifier. Such an algorithm may not exist for several reasons. The question of when two terms have a unifier may be undecidable, in which case one looks for a *semi*-algorithm that may not always terminate. It can happen that two unifiable terms fail to have a most general unifier, in which case one might look for a *complete* set Σ of unifiers such that every unifier is less general than some element of Σ. Finally, Σ might necessarily be infinite, in which case, one might look for an algorithm to enumerate Σ or for a finite description of Σ. All three of these problems arise in Example 5 below.

A general introduction to unification and its applications has been given by Knight [KNIG89]. A comprehensive technical survey of unification theory has been given by Siekmann [SIEK89].

Discussion of computational complexity for unification algorithms must be based on an appropriate measure of term *size*. If size is just number of symbols, then the size of the common instance t in Exercise 2.4A is exponential in the size of the terms to be unified, as is the size of the unifier σ^*. Efficient unification can avoid this exponential explosion by using graph implementations in which repeated occurrences of a subterm are stored only once.

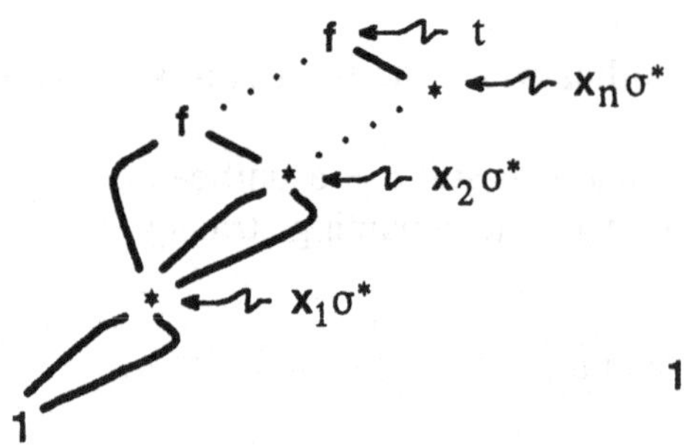

t with structure-sharing

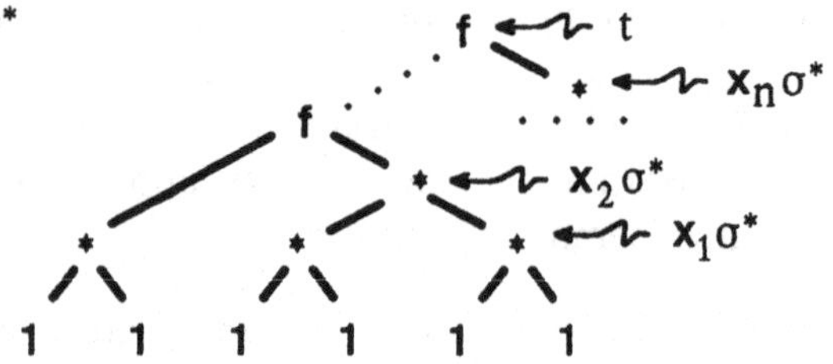

t without structure-sharing

The above term $t = r\sigma^*$, for example, may be implemented by either of the above finite graphs. Since $\mathbf{x}_2\sigma^* = \mathbf{x}_1\sigma^*{*}\mathbf{x}_1\sigma^*$, the term $\mathbf{x}_1\sigma^*$ occurs in t in at least three different ways. The needed structure-sharing may be achieved during computation of σ^*. Specifically, σ is iteratively converted to σ^* as follows: whenever $v\sigma$ contains a variable u in the domain of σ, pointers to u in the graph of $v\sigma$ are moved so that they point to $u\sigma$ instead.

Many unification algorithms produce unifiers in a form similar to σ^*. Jaffar's algorithm returns σ itself as an implementation of σ^* [JAFF84]. The algorithm of Martelli and Montanari builds a "triangular form" substitution that facilitates computation of σ^* [MART82]. Robinson's algorithm returns an indicated composition of the form $(t_1/v_1) \ldots (t_n/v_n)$ [ROBI79, CORB83], but this composition can be written as σ^*, where $\sigma = (t_1/v_1) + \ldots + (t_n/v_n)$. In this case, $\sigma^* = \sigma^n$.

The above discussion suggests that structure sharing in graph implementations may be understood abstractly in terms of iteration; moreover, this understanding is compatible with the design of well-known unification algorithms.

MOTIVATING EXAMPLES

2.1. *Quotient Instantiation*. Equational identities provide a natural source of term congruences. Any set of identities closed under logical consequence *generates* a term congruence in an obvious way. For example, the associative law of elementary arithmetic generates a term congruence, and the resulting quotient system is referred to as *associative instantiation*. For any set E of equations, the quotient system *induced* by E is referred to as E-*instantiation*.

2.2. *String Instantiation* [SIEK75].

A. The terms are nonempty finite strings over an arbitrary alphabet A that contains a distinguished subset V of *variables*. Elements of A \ V are referred to as *constants*. By abuse of notation, each a in A is identified with the one element string 'a.' Substitutions replace variables with strings. Given a string s and substitution σ, $s\sigma$ is the string obtained by replacing each variable v in s with the string $v\sigma$ in the obvious way.

B. Example 2.2A is isomorphic to the quotient action obtained as follows: Expand the language of semigroups to include constant symbols from the alphabet A \ V. Starting with first-order term instantiation, take the quotient action that results from the term congruence generated by the associative law. To see the isomorphism, just interpret the monoid operation as string concatenation.

2.3. *Predicate Calculus*. The syntactic objects are the terms and formulas of full first-order predicate calculus.

A. *Naive Instantiation*. As with first-order terms, instantiation is by straight plugging in. However, $\phi\sigma$ is undefined whenever v occurs bound in ϕ and vσ is not a variable. Semantic constraints are not considered, so, for example, **(∃ x) (x < x)** is an instance of **(∃ x) (x < y)**. (As in this case, examples of syntactic objects are displayed using boldfaced type.)

B. *Semantically Restricted Instantiation*. Disallow semantically wrong instantiations such as the one above; instantiation is still a partial monoid action.

C. *Alpha Instantiation*. Starting with the above partial instantiation, take the quotient system induced by α–conversion, that is, by the rule

(∃ u**)** ϕ(u) iff **(∃** v**)** ϕ(v), provided v is free for u in ϕ(u).

Notice that instantiation is always defined in the resulting quotient system. (As in this example, schemas are written by embedding non–bold expressions such as 'u' or 'ϕ(u)' in a displayed syntactic object; the intent is that such *metaexpressions* are to be replaced with syntactic objects that they may denote.)

2.4. *Lambda Instantiation* [cf JENS73, HUET75]. Start with the terms of lambda calculus. These consist of *variables* **x**, **y**, **f**, **g**, ..., *constant symbols* **X**, **Y**, **F**, **G**, ..., *applications* of the form s(t), and λ*-abstractions* of the form **(λ**v**.**t**)**. Application associates to the left and binds more tightly than λ-abstraction. For instantiation, begin with straight plugging in, restricted to avoid replacement of bound variables. Then take the quotient system induced by the α– and β–conversion rules.

2.5. *Second Order Instantiation* [GOLD81]. The terms of second–order logic may be regarded as a sublanguage of lambda calculus. One arbitrarily separates variables into two infinite sets of *function* and *individual* variables, and then uses only **λ**-free terms where variables occur appropriately. Each such lambda term may be viewed as (the Curried form of) a second–order term. We now obtain a subsystem of Example 2.4 by using only substitutions that always instantiate second–order terms to second–order instances. Thus, if σ is the substitution **(λx.**t**)/f**, that replaces the function variable **f** with a term of the form **(λx.**t**)**. Then β–conversion allows a term of the form **f(**s**)**σ to be identified with the term t(s/**x**) obtained by replacing free occurrences of **x** with s. This is still second–order, provided t and s are. The unification problem for second–order instantiation and, hence, for lambda instantiation, is undecidable.

SECTION 3
GENERAL APPROACHES TO INSTANTIATION

This section introduces a theory of *instantiation systems* and summarizes related approaches to instantiation based on tree systems, order-sorted algebras, algebraic theories, and composita. Tree systems play a major role in Section 6. Key concepts from order-sorted algebras turn up occasionally in later sections. The material on algebraic theories and composita is included primarily to show how the present approach relates to the existing literature, and is not necessary for later sections.

INSTANTIATION SYSTEMS

The following theory formalizes the concept of a *(syntactic) variable*. It is based on the idea that variables form a minimal discriminating set for a faithful monoid action. Variables are introduced axiomatically because minimal discriminating sets are not unique (see Exercise 3.3 below).

The first four axioms recast the theory of faithful unital monoid actions. The fifth guarantees that there are enough substitutions to form a reasonable partial-function algebra. The sixth rules out pathological examples; its statement allows for the possibility that not all variables are of the same type. The last axiom guarantees that no term or substitution can contain all variables of a given type.

The equivalent axiomatization by Schmidt-Schauss and Siekmann [SCHM88] replaces Axiom 7 with the assumption that var(t) is finite; their definition of *var*(t) is identical to the one given below. In both developments, the first significant result is that var(t) is the smallest discriminating set for t. This establishes the equivalence of the two theories.

Consider a set O of *terms*, a set $V \subseteq O$ of *variables*, a set S of finite partial functions from V to O called *substitutions*, and a function $*$ from $O \times S$ to O called *instantiation*; we write $t\sigma$ in favor of $*(t, \sigma)$. Let r, s, t, ... vary over terms; let u, v, w, ... vary over variables; let ρ, σ, τ, ... vary over substitutions.

Let ϵ be the empty function. Define $s \leq t$ to mean that $s\sigma = t$, for some $\sigma \in S$; s and t are *equivalent* iff $s \leq t$ and $t \leq s$. Let *dom*(σ) be the domain of σ. Let $\sigma\tau$, the *composition* of σ and τ, be the function with minimal domain such that $v(\sigma\tau) = (v\sigma)\tau$, for all $v \in V$. For any set X, $\sigma|X$ has domain $dom(\sigma) \cap X$ and is the restriction of σ to X. We say X *discriminates for* a term t iff $t\sigma = t\tau$, whenever $\sigma|X = \tau|X$.

The four-tuple (O, V, S, $*$) is an *instantiation system* iff the following axioms are satisfied, for all t, v, σ, and τ:

1 (*closure*). $\sigma\tau \in S$.
2 (*associativity*). $t(\sigma\tau) = (t\sigma)\tau$.
3 (*identity*). $\epsilon \in S$ and $t\epsilon = t$.
4 (*representation*). $v\sigma = \sigma(v)$, if $v \in dom(\sigma)$;
$v\sigma = v$ iff $v \notin dom(\sigma)$.
5 (*restriction*). $\sigma|X \in S$, for each $X \subseteq V$.
6 (ω-*repleteness*). $\{u \mid u \leq v$ and $v \leq u\}$ is infinite.
7 (ω-*discrimination*). X discriminates for t, for some finite $X \subseteq V$.

From Section 2, we see easily that composition of substitutions is a monoid operation, and that instantiation is a faithful, unital monoid action represented as an action by substitutions on V.

First-order term instantiation and Example 2.2A may be expanded to instantiation systems by taking V to be the set of all variables. Examples 2.1, 2.2B, 2.3C, and 2.4 may be expanded to instantiation systems by taking V to be the set of all equivalence classes of the form {v}, where v is a variable. Examples 2.3A and 2.3B do not expand to instantiation systems, because instantiation may be undefined. Finally, Example 2.5 does not immediately expand to an instantiation system, because function variables are not second-order terms. An alternate set of "variables" for Example 2.5 is given in Exercise 3.3 below.

Two terms s and t are *equivalent* iff $s \leq t$ and $t \leq s$. The ω-repleteness axiom just says that every variable is equivalent to infinitely many other variables. An instantiation system is *untyped* iff every term is an instance of every variable; it *has untyped variables* iff every variable is an instance of every variable. Example 2.5 (second-order instantiation) has two variable types, Example 2.3 (predicate calculus) has untyped variables, and Examples 2.1, 2.2, and 2.4 are untyped. In general, the instance relation restricted to variables can be arbitrarily complex (see Exercise 3.12B below).

If f, g are partial functions from V to O, then f + g is the union of f and g, provided $dom(f) \cap dom(g) = \emptyset$; we shall soon show that $\sigma + \tau$ is a substitution, whenever $\sigma + \tau$ is defined.

Exercise 3.1. In any instantiation system,

A. $\sigma + \tau = \tau + \sigma$.
B. $(\sigma + \tau) + \eta = \sigma + (\tau + \eta)$.
C. $dom(\sigma + \tau) = dom(\sigma) \cup dom(\tau)$.
D. $dom(\sigma|X) \subseteq X$.
E. $(\sigma|X)|Y = \sigma|(X \cap Y)$.

F. $\sigma = \sigma|X$, if $\mathrm{dom}(\sigma) \subseteq X$.
G. $(\sigma + \tau)|X = \sigma|X$, if $\sigma + \tau$ is defined and $X \cap \mathrm{dom}(\tau) = \emptyset$.
H. $\sigma\tau = (\sigma\tau)|\mathrm{dom}(\sigma) + \tau|(V \setminus \mathrm{dom}(\sigma))$.
I. $(\sigma + \tau)\eta = (\sigma\eta)|\mathrm{dom}(\sigma) + (\tau\eta)|(V \setminus \mathrm{dom}(\sigma))$, if $\sigma + \tau$ is defined.

(These properties follow directly from the fact that substitutions are partial functions from V to O.)

If $v \leq t$, then there is a unique substitution σ such that $v\sigma = t$ and $\mathrm{dom}(\sigma) \subseteq \{v\}$. We refer to this as the *quotient* substitution t/v.† If $t \neq v$, then t/v is just the one-element function that maps v to t.

Exercise 3.2. Assume $v \leq t$, then
A. $v(t/v) = t$.
B. $u(t/v) = u$, if $u \neq v$.
C. $(u/v)(v/u) = v/u$, if $u \leq v$ and $v \leq u$.
D. $(v\sigma)/v = \sigma|\{v\}$.
E. $((t/v)\sigma)|\{v\} = (t\sigma)/v$.
F. $\sigma = v_1\sigma/v_1 + \dots + v_n\sigma/v_n$, if $\mathrm{dom}(\sigma) = \{v_1, \dots, v_n\}$.

(To prove assertion C, for example, show that $w(u/v)(v/u) = w(v/u)$, for any $w \in V$, by considering the cases $w = v$, $w = u$, and $w \notin \{u, v\}$.)

Exercise 3.3. Regarding Example 2.5, let **H** be a binary function constant. Let $\mathbf{C}_1$, $\mathbf{D}_1$, $\mathbf{C}_2$, $\mathbf{D}_2$, ... be an enumeration of the individual constants. Let V_1 be the set of all individual variables. Let V_2 be the set of all terms of the form $\mathbf{H}(f(\mathbf{C}_1, \dots, \mathbf{C}_n), f(\mathbf{D}_1, \dots, \mathbf{D}_n))$, where f is any n-ary function variable. Let $V = V_1 \cup V_2$.

A. For any term t, t is uniquely characterized, up to a renaming of variables, as the least-general term which is a common generalization of $t(\mathbf{C}_1/\mathbf{x}_1 + \dots + \mathbf{C}_n/\mathbf{x}_n)$ and $t(\mathbf{D}_1/\mathbf{x}_1 + \dots + \mathbf{D}_n/\mathbf{x}_n)$.
B. $[\lambda \mathbf{x}_1, \dots, \mathbf{x}_n.t]$ is recoverable from $\mathbf{H}(f(\mathbf{C}_1, \dots, \mathbf{C}_n), f(\mathbf{D}_1, \dots, \mathbf{D}_n))([\lambda \mathbf{x}_1, \dots, \mathbf{x}_n.t]/f)$.
C. V is a minimal discriminating set that may be used to recast Example 2.5 as an instantiation system.‡
D. The "variable" $\mathbf{H}(f(\mathbf{C}_1, \dots, \mathbf{C}_n), f(\mathbf{D}_1, \dots, \mathbf{D}_n))$ is an instance of the nonvariable $\mathbf{H}(f(\mathbf{C}_1, \dots, \mathbf{C}_n), f(\mathbf{x}_1, \dots, \mathbf{x}_n))$.

† The notation has been chosen by analogy with the natural numbers — multiplication makes things bigger and the cancellation law in Exercise 3.2A looks familiar.

‡ Clearly, there is nothing unique about V.

A substitution η is *variable-valued* iff $v\eta \in V$, for all $v \in V$. If η is variable-valued, then *cdm*$(\eta) = \{v\eta \mid v \in \mathrm{dom}(\eta)\}$. A *renaming* substitution is a variable-valued substitution ρ that, as a function, is invertible, and whose inverse $\rho^{\vee}$ is also a substitution. In this case, we refer to $\rho^{\vee}$ as the *functional inverse* of ρ. A substitution σ is *invertible (as a substitution)* iff there is a substitution τ such that $\sigma\tau = \tau\sigma = \epsilon$. In general, invertible substitutions needn't be variable-valued (see Example 4.1B in the next section).

Exercise 3.4. Assume ρ is a renaming substitution.

A. If σ is invertible then its (*substitutional*) *inverse* is unique.

B. $(\rho\rho^{\vee})|\mathrm{dom}(\rho) = (\rho^{\vee}\rho)|\mathrm{dom}(\rho^{\vee}) = \epsilon$; ρ is invertible as a substitution iff ρ is a permutation iff $\mathrm{dom}(\rho) = \mathrm{cdm}(\rho)$.

C. If $\mathrm{dom}(\rho)$ is disjoint from $\mathrm{cdm}(\rho)$, then $\rho + \rho^{\vee}$ is its own inverse.

Lemma 3.5. Let A, B be finite subsets of V. Then there is a renaming substitution ρ such that $\mathrm{dom}(\rho) = A$ and $\mathrm{cdm}(\rho) \cap B = \emptyset$.

Proof. Let $A = \{v_1, ..., v_n\}$. Using ω-repleteness, we may choose, for each v_i, an equivalent u_i, with $u_i \notin A \cup B \cup \{u_1, ..., u_{i-1}\}$. Let $\rho = (u_1/v_1) \ldots (u_n/v_n)$. Clearly, ρ is a renaming substitution with functional inverse $\rho^{\vee} = (v_1/u_1) \ldots (v_n/u_n)$. Moreover, $\mathrm{cdm}(\rho)$ is disjoint from B. □

Lemma 3.6.

A. X discriminates for t iff $t\sigma = t(\sigma|X)$, for all σ.

B. If ρ is variable-valued and D discriminates for t, then $\{v\rho \mid v \in D\}$ discriminates for $t\rho$.

C. $\sigma + \tau$ is a substitution, if $\sigma + \tau$ is defined.

Proof. To prove A, first suppose that X discriminates for t. Pick σ. Then $\sigma|X = (\sigma|X)|X$, so that $t\sigma = t(\sigma|X)$. Conversely, suppose $t\sigma = t(\sigma|X)$, for all σ. Assume σ, τ are such that $\sigma|X = \tau|X$. Then $t\sigma = t(\sigma|X) = t(\tau|X) = t\tau$. Hence X discriminates for t. To prove Assertion B, let $Y = \{v\rho \mid v \in D\}$. To see that Y discriminates for $t\rho$, suppose $\sigma|Y = \tau|Y$. Then $(\rho\sigma)|D = (\rho\tau)|D$. Consequently, $t\rho\sigma = t\rho\tau$, since D discriminates for t. We have just shown that Y discriminates for $t\rho$. For Assertion C, assume $\sigma + \tau$ is defined. Let X be a finite set of variables that discriminates for $v\sigma$, if $v \in \mathrm{dom}(\sigma)$, and for $v\tau$, if $v \in \mathrm{dom}(\tau)$, for all v. Let η be a renaming such that $\mathrm{dom}(\eta) = X$ and $\mathrm{cdm}(\eta)$ is disjoint from $X \cup \mathrm{dom}(\tau)$. Then, $\sigma + \tau = (((\sigma\eta)|\mathrm{dom}(\sigma))\tau\eta^{\vee})|(\mathrm{dom}(\sigma) \cup \mathrm{dom}(\tau))$. □

Let *var*$(t) = \{v \mid t(s/v) \neq t$, for some $s\}$. Notice that $\mathrm{var}(v) = \{v\}$, for any $v \in V$. In Example 2.3C, var(t) is just the set of variables that occur free in the term or formula t. We can now extend the cdm operator by letting *cdm*$(\sigma) = \bigcup\{\mathrm{var}(v\sigma) \mid v \in \mathrm{dom}(\sigma)\}$. Thus, $\mathrm{cdm}(\sigma)$ coincides with the range of the function σ iff σ is variable-valued.

Theorem 3.7. X discriminates for t iff $\mathrm{var}(t) \subseteq X$. In other words, var(t) is the smallest set which discriminates for t.†

Proof. First, if X discriminates for t, then $\mathrm{var}(t) \subseteq X$: If not, then choose $v \in \mathrm{var}(t) \setminus X$. Choose s so that $t(s/v) \neq t$. Since $(s/v)|X = \epsilon$, we have $t(s/v|X) = t \neq t(s/v)$. But then X doesn't discriminate for t, by Lemma 3.6A.

Now let D be any minimal discriminating set for t. Suppose $v \in D \setminus \mathrm{var}(t)$. Let $C = D \setminus \{v\}$. By assumption, C fails to discriminate for t, so we may choose σ so that $\mathrm{dom}(\sigma) = D$ and $t\sigma \neq t(\sigma|C)$. Let A be a finite set that discriminates for $v\sigma$. Let B be a finite set that discriminates for each term in $\{u\sigma \mid u \in C\}$. Using Lemma 3.5, let ρ be a renaming substitution such that $\mathrm{dom}(\rho) =^{\sigma} A$ and $\mathrm{cdm}(\rho)$ is disjoint from $A \cup B \cup D$. Notice that $\mathrm{cdm}(\rho)$ discriminates for $v\sigma\rho$, by Lemma 3.6B. Finally, let $\tau = (v\sigma\rho)/v + \sigma|C$. The definition of τ may be summarized as follows:

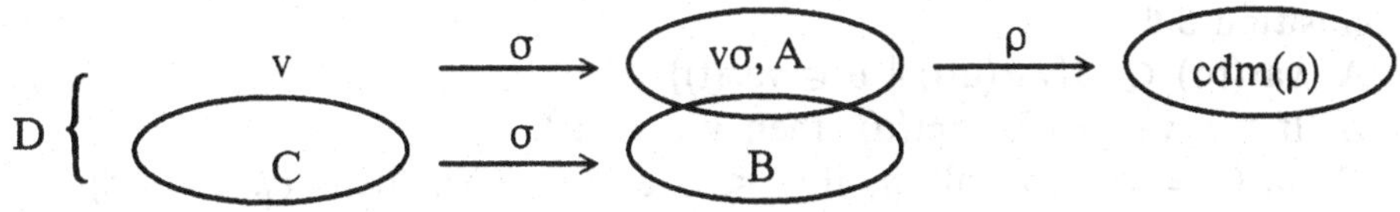

1. $(\tau\rho^{\vee})|D = \sigma$: $v\tau\rho^{\vee} = v\sigma\rho\rho^{\vee} = v\sigma$. If $u \in C$, then $u\tau\rho^{\vee} = u(\sigma|C)\rho^{\vee} = u\sigma\rho^{\vee} = u\sigma$, since $\mathrm{dom}(\rho^{\vee})$ is disjoint from B, and B discriminates for $u\sigma$.
2. $((\tau|C)\rho^{\vee})|D = \sigma|C$: $v(\tau|C)\rho^{\vee} = v\rho^{\vee} = v = v(\sigma|C)$, since neither C nor $\mathrm{dom}(\rho^{\vee})$ contain v. If $u \in C$, then $u(\tau|C)\rho^{\vee} = u\sigma\rho^{\vee} = u\sigma = u(\sigma|C)$.
3. $t\tau \neq t(\tau|C)$: Otherwise, $t\sigma = t((\tau\rho^{\vee})|D) = t\tau\rho^{\vee} = t(\tau|C)\rho^{\vee} = t(((\tau|C)\rho^{\vee})|D) = t(\sigma|C)$, by Step 1, choice of D, the counter hypothesis, choice of D, and Step 2. This contradicts our choice of σ.
4. $\tau|D = (\tau|\{v\})(\tau|C)$: $v(\tau|\{v\})(\tau|C) = (v\sigma\rho)(\tau|C) = (v\sigma\rho)((\tau|C)|\mathrm{cdm}(\rho)) = v\sigma\rho = v\tau$, by choice of τ, that $\mathrm{cdm}(\rho)$ discriminates for $v\sigma\rho$, that C is disjoint from $\mathrm{cdm}(\rho)$, and choice of τ. If $u \in C$, then $u(\tau|\{v\})(\tau|C) = u(\tau|C) = u\tau$, since $u \neq v$ and $u \in C$.

Finally, $t\tau = t(\tau|D) = t(\tau|\{v\})(\tau|C) = t((v\sigma\rho)/v)(\tau|C) = t(\tau|C)$, by choice of D, Step 4, choice of τ, and the fact that $v \notin \mathrm{var}(t)$. This contradicts Step 3. □

Corollary 3.8.

A. $t\sigma = t(\sigma|\mathrm{var}(t))$, for all σ.

B. $t\sigma = t$, if $\mathrm{var}(t) \cap \mathrm{dom}(\sigma) = \emptyset$.

---

† This, together with [SCHM88, Corollary 4.11], establishes the equivalence of instantiation systems and unification algebras.

C. $t(\sigma + \tau) = t\sigma$, if $\mathrm{dom}(\tau) \cap \mathrm{var}(t) = \emptyset$ and $\sigma + \tau$ is defined.
D. $\sigma + \tau = \sigma\tau$, if $\mathrm{cdm}(\sigma) \cap \mathrm{dom}(\tau) = \emptyset$ and $\sigma + \tau$ is defined.
E. If $\mathrm{dom}(\sigma) \cap \mathrm{cdm}(\sigma) = \emptyset$, then $\sigma^2 = \sigma$.

Proof. Assertion A follows directly from Lemma 3.6A and Theorem 3.7. Assertion B is a special case of Assertion C, which, in turn, follows directly from Assertion A and Exercise 3.1G, with $X = \mathrm{var}(t)$.

Assertion D is proved by considering an arbitrary v in $\mathrm{dom}(\sigma + \tau)$: If $v \in \mathrm{dom}(\sigma)$, then $v(\sigma + \tau) = v\sigma = (v\sigma)\tau$, since $\mathrm{var}(v\sigma) \subseteq \mathrm{cdm}(\sigma)$ and $\mathrm{cdm}(\sigma) \cap \mathrm{dom}(\tau) = \emptyset$. Similarly, if $v \in \mathrm{dom}(\tau)$, then $v(\sigma + \tau) = v\tau = v\sigma\tau$, since $v \notin \mathrm{dom}(\sigma)$. Assertion E is similarly straightforward. □

A *weakening for* t is a variable-valued substitution η such that $u\eta = v\eta$ implies $u = v$, for all $u, v \in \mathrm{var}(t)$.

Proposition 3.9.
A. $\mathrm{var}(t\sigma) \subseteq \bigcup\{\mathrm{var}(u\sigma) \mid u \in \mathrm{var}(t)\}$.
B. If $v \in \mathrm{var}(t\sigma) \setminus \mathrm{cdm}(\sigma)$, then $v \in \mathrm{var}(t)$.
C. $\mathrm{var}(t) = \{v \mid \text{for all } w, \text{ if } v \leq w \text{ and } v \not\approx w, \text{ then } t(w/v) \not\approx t\}$.
D. A variable-valued substitution η is a weakening for t iff $\eta|\mathrm{var}(t)$ is one-to-one and $\mathrm{cdm}(\eta|\mathrm{var}(t)) \cap \mathrm{var}(t) \subseteq \mathrm{dom}(\eta)$.

Proof. Assertion A is proved as follows: Pick $v \in \mathrm{var}(t\sigma)$. Choose s so that $t\sigma(s/v) \not\approx t\sigma$. Since var(t) discriminates for t, we may choose $u \in \mathrm{var}(t)$ so that $u\sigma(s/v) \not\approx u\sigma$. This shows that $v \in \mathrm{var}(u\sigma)$, as required. Assertion B follows immediately from Assertion A.

Regarding Assertion C, it suffices to pick $v \in \mathrm{var}(t)$, assume $v \leq w$ and $v \not\approx w$, and show $t(w/v) \not\approx t$. If $w \in \mathrm{var}(t)$, then $\mathrm{var}(t(w/v)) \subseteq \mathrm{var}(t) \setminus \{v\}$, by Assertion A, so that $t(w/v) \not\approx t$. Similarly, if $w \notin \mathrm{var}(t)$, then $\mathrm{var}(t(w/v)) \subseteq (\mathrm{var}(t) \setminus \{v\}) \cup \{w\}$, so again, $t(w/v) \not\approx t$.

Regarding Assertion D, by definition, η is a weakening for t iff $\eta|\mathrm{var}(t)$ is one-to-one and for all $u, v \in \mathrm{var}(t)$, $u \in \mathrm{dom}(\eta)$ and $v \notin \mathrm{dom}(\eta)$ implies $u\eta \not\approx v\eta$. This latter condition simplifies to $\mathrm{cdm}(\eta|\mathrm{var}(t)) \cap \mathrm{var}(t) \subseteq \mathrm{dom}(\eta)$. □

Let $\ll$ be an arbitrary (but fixed) well ordering of V. For any term s and terms $t_1, \ldots, t_n$, we define $s\langle t_1, \ldots, t_n\rangle$ to be $s\sigma$, where $\{v_1, \ldots, v_n\} = \mathrm{var}(s)$, $v_1 \ll \ldots \ll v_n$, and $\sigma = t_1/v_1 + \ldots + t_n/v_n$, provided this is defined.†

† Thus, we consider a term to be a function of its variables and use substitutions as variable-bindings.

Proposition 3.10. If $s\langle t_1, ..., t_n\rangle$ is defined, then

A. $s\langle t_1, ..., t_n\rangle\sigma = s\langle t_1\sigma, ..., t_n\sigma\rangle$.

B. $var(s\langle t_1, ..., t_n\rangle) \subseteq var(t_1) \cup ... \cup var(t_n)$.

Proof. To prove Assertion A, let $\tau = t_1/v_1 + ... + t_n/v_n$, where $v_1 \ll ... \ll v_n$, with $var(s) = \{v_1, ..., v_n\}$. Then $(\tau\sigma)|var(s) = t_1\sigma/v_1 + ... + t_n\sigma/v_n$, as can be seen by applying both sides to an arbitrary v_i. Consequently, $s\langle t_1, ..., t_n\rangle\sigma = s\tau\sigma = s((\tau\sigma)|var(s)) = s\langle t_1\sigma, ..., t_n\sigma\rangle$. Assertion B is just a recasting of Proposition 3.9A. □

TREE INSTANTIATION SYSTEMS

Courcelle has given a comprehensive introduction to trees and related instantiation issues [COUR83]. We will be primarily interested in *finite* trees. Our definition of *trees* omits the traditional notion of a *ranked alphabet*. Here, the role of *rank* functions is subsumed by more general techniques inspired by order-sorted algebras. Moreover, variable-arity trees are needed in Example 6.5.

For any nonempty set A, the class *TR* of all (possibly infinite) *trees with labels in* A may be abstractly characterized via a structure of the form TR#(A) = (TR, A, ⟪_, ...⟫), where

1. ⟪_, ...⟫ is a bijection from $A \times \bigcup\{TR^n \mid n \geq 0\}$ onto TR.
2. TR#(A) is maximal among all similar structures having the following property: Equality is the largest equivalence relation E such that whenever $⟪a, r_1, ..., r_n⟫ \mathrel{E} ⟪b, s_1, ..., s_n⟫$, we have $a = b$ and $r_i \mathrel{E} s_i$, for each i.

An *atom* is a tree of the form ⟪a⟫. The class of *finite* trees is the smallest subset *FTR* of TR that contains every atom and is closed under ⟪_, ...⟫.† For finite trees, one may identify $⟪a, r_1, ..., r_n⟫$ with the sequence $\langle a, r_1, ..., r_n\rangle$. For any tree $r = ⟪a, r_1, ..., r_n⟫$, the trees $r_1, ..., r_n$ are the *immediate subtrees* of r, and we define $r\hat{\ }j$ to be the subtree r_j, for $j = 1, ..., n$. A sequence $\langle j_1, ..., j_p\rangle$ is an *occurrence* of r in t iff $r = t\hat{\ }j_1\hat{\ }...\hat{\ }j_p$. The set of all *subtrees* of r is the smallest set ST(r) that contains r and is closed under the formation of immediate subtrees. The class *RTR* of all *rational* trees consists of those trees r such that ST(r) is finite.

† This use of the word "finite" is both well-established and unfortunate. It would be better to refer to these trees as *well-founded*, because rational trees are also finite in an obvious sense.

If V is infinite and disjoint from A, TR#(A ∪ V) can be made into an untyped instantiation system *TR#*(A, V) as follows: Identify V with {⟪v⟫ | v ∈ V}, which we take to be the set of variables. Substitutions are finite partial functions from variables to trees. Given a tree t and substitution σ, tσ is obtained from t by straight "plugging in." That is, vσ is either σ(v) or v, depending on whether v ∈ dom(σ); if a ∈ A, then $\langle\langle a, r_1, ..., r_n\rangle\rangle\sigma = \langle\langle a, r_1\sigma, ..., r_n\sigma\rangle\rangle$. The instantiation subsystems *FTR#*(A, V) and *RTR#*(A, V) are obtained by restricting TR#(A, V) to finite and rational trees, respectively.

Exercise 3.11.

A. FTR#(A, V), RTR#(A, V), and TR#(A, V) are instantiation systems.

B. If A and V are countably infinite, then FTR#(A, V) is isomorphic with first-order term instantiation. This is true in spite of the fact that elements of A must be multiply interpreted as individual constants and as n-ary function symbols, for all n. Assume, for simplicity, that V is just the set of all first-order variables.

C. If $s = t\hat{\ }j_1\hat{\ }...\hat{\ }j_p$ and $v \notin var(t)$, then there is a unique tree r such that $t = r(s/v)$. In particular, if $t = t\hat{\ }j_1\hat{\ }...\hat{\ }j_p$, then $t = r(t/v)$.

ORDER-SORTED INSTANTIATION

The *order-sorted* instantiation systems of Exercise 3.12A give the instantiation mechanism used for term algebras of the form $\mathcal{T}_\Sigma(V)$, as defined by Meseguer, Goguen, and Smolka [GOGU87, MESE87], with the following two differences: Their definitions assume monotonicity, but not ω-repleteness. Our v-regularity condition is a weakening of their regularity condition. Exercise 3.12C touches on a limitation of order-sorted systems that is not related to these differences (see Example 5.2A and Exercises 5.6C and 6.6B).

An *order-sorted signature* is a triple of the form $(\mathcal{T}, \subseteq, \mathcal{S})$, where $\mathcal{T}$ is an arbitrary set of *sorts,* ⊆ is a partial order on $\mathcal{T}$, and $\mathcal{S}$ is a function whose domain consists of the variables, constant symbols and function symbols of predicate calculus, where $\mathcal{S}(b) \subseteq \mathcal{T}$, for each variable or constant symbol b, and where $\mathcal{S}(f) \subseteq \mathcal{T}^{n+1}$, for each n-ary function symbol f. Let T, T_1, T_2, ... vary over elements of $\mathcal{T}$. We write $f : T_1 \times ... \times T_n \to T$ as an abbreviation for $\langle T_1, ..., T_n, T\rangle \in \mathcal{S}(f)$. Inductively define an *of-type* relation between terms and types as follows: If $T \in \mathcal{S}(b)$, then b is of type T; if t_i is of type T_i, for $i = 1, ..., n$, and $f : T_1 \times ... \times T_n \to T$, then $f(t_1, ..., t_n)$ is of type T; finally, if t is of type T_1 and $T_1 \subseteq T_2$, then t is of type T_2. A term is *well-typed* iff it is of type T, for some T. A substitution σ is *type-preserving* iff vσ is of type T whenever v is. A trivial induction shows that if σ is type preserving, then tσ is of type T whenever t is, for any term t.

An order-sorted signature is:

ω-replete iff for each variable v, there are infinitely many variables w such that $\mathcal{I}(w) = \mathcal{I}(v)$.

v-regular iff each variable v is of type T_v, for some (unique) minimum type T_v.

monotone iff whenever $S_1 \times \ldots \times S_n \text{->} S$ and $T_1 \times \ldots \times T_n \text{->} T$ are both signatures of f, with $S_i \subseteq T_i$, for each i, then $S \subseteq T$.

Exercise 3.12. Let $(\mathcal{T}, \subseteq, \mathcal{I})$ be an ω-replete order-sorted signature.

A. The well-typed terms, variables, type-preserving substitutions, and inherited instantiation operation form an instantiation system, which we refer to as the *associated order-sorted* instantiation system.

B. If the signature is v-regular, then $u \leq v$ iff $T_v \subseteq T_u$, so that type structure is recoverable from the instance relation. Thus, the instance relation on variables can be arbitrarily complex.

C. Assume the signature is v-regular and that whenever $t \in T_1 \cap T_2$, there exists w such that $w \leq t$ and $w \in T_1 \cap T_2$. Then terms are *restrictable* in the sense that if $t\sigma \in T$, then there exists η such that η is a weakening for t, $t\eta \in T$ and $t\eta \leq t\sigma$ [cf. WALT86, Sect. 3].

D. Consider the order-sorted instantiation system with signature given by $\mathcal{T} = \{T_1, T_2, T\}$; $T \subseteq T_1$, $T \subseteq T_2$; $\mathbf{x}_1, \mathbf{y}_1, \ldots \in T_1$; $\mathbf{x}_2, \mathbf{y}_2, \ldots \in T_2$; $\mathbf{x}, \mathbf{y}, \ldots \in T$. The substitution $\sigma = \mathbf{x}/\mathbf{x}_1 + \mathbf{x}/\mathbf{x}_2$ is a most general unifier of $\mathbf{x}_1$ and $\mathbf{x}_2$. If σ′ is any other most-general unifier of $\mathbf{x}_1$ and $\mathbf{x}_2$, then $\sigma \neq \sigma'\eta$, for all η, and we cannot have $\text{cdm}(\sigma') \subseteq \{\mathbf{x}_1, \mathbf{x}_2\}$.

OTHER SYSTEMS OF HISTORICAL INTEREST

The observation that variables form a discriminating set for a faithful monoid action is due to Nerode, as is a precursor to the structure theorem in Section 6; his theory of composita is a deliberate generalization of first-order term instantiation [NERO59]. The notion of iteration presented in Section 4 is essentially due to Elgot; his work was primarily motivated by the fact that algebraic theories (and hence, instantiation systems) may be used to model a certain class of infinite state machines. Briefly, substitutions are like state transformations, and terms are like system-initializations [ELGO75].

An algebraic theory may be viewed as an implementation of an untyped instantiation system: each substitution σ is implemented as a triple ⟨m, σ, n⟩, where m and n are estimates of the domain and codomain of σ, respectively. These triples are the mappings of a category.

Composita

A *compositum* [NERO59] is a structure of the form $\mathcal{C} = (O, V, S)$, where O contains at least two elements, V is a nonempty subset of O, and S is a set of mappings from O to O such that
1) S contains the identity map and is closed under composition,
2) every map from V to O has a unique extension in S.

Exercise 3.13.
A. A compositum is a representation by transformations (in the sense of Section 2) of a monoid action.
B. Composita are untyped, in the sense that $u \leq t$, for all $u \in V$ and $t \in O$.
C. Let $\mathcal{A}$ be an untyped instantiation system. Let $\mathcal{A}^c$ be the structure (O, V, S^c), where S^c is the set of all functions α from O to O such that for some σ, $\alpha(t) = t\sigma$, for all $t \in O$. Then $\mathcal{A}^c$ is a compositum.
D. The mapping $\mathcal{A} \mapsto \mathcal{A}^c$ is one-to-one.
E. In any compositum, we define X to be a *discriminating* set for t iff $X \subseteq V$ and, for all α, β, if $\alpha|X = \beta|X$, then $\alpha(t) = \beta(t)$. Let $\mathcal{C}$ be a compositum such that V is infinite and every term has a finite discriminating set. Then $\mathcal{C}$ is of the form $\mathcal{A}^c$.

Algebraic Theories

For any instantiation system $\mathcal{A}$, we define a corresponding category $\mathcal{A}^C$ as follows: Let $\langle v_1, v_2, \ldots \rangle$ be a fixed enumeration of V. For each natural number n, let $V_n = \{v_1, \ldots, v_n\}$. For each m and n, let $T^m{}_n$ be the set of all triples $\langle m, \sigma, n\rangle$ where $\mathrm{dom}(\sigma) \subseteq V_m$ and $\mathrm{var}(v\sigma) \subseteq V_n$, for all $v \in V_m$. These triples are the *morphisms* of $\mathcal{A}^C$; *composition* of morphisms is given by

$$\langle m, \sigma, p\rangle \circ \langle p, \tau, n\rangle = \langle m, (\sigma\tau)|V_m, n\rangle.$$

The *objects* of $\mathcal{A}^C$ are the natural numbers. Each $T^m{}_n$ is the set of all *mappings from* m *to* n. For each n, $\langle n, \epsilon, n\rangle$ is the identity mapping in $T^n{}_n$.

As defined by Elgot [ELGO75], an *algebraic theory* consists of a small category whose objects are the natural numbers, together with a doubly indexed family of *coordinate* morphisms, $\nu_{im} \in T^1{}_m$, where for each m, the mapping $\chi \mapsto \langle \nu_{1m}\circ\chi, \ldots, \nu_{mm}\circ\chi\rangle$ is a bijection from $T^m{}_n$ to $(T^1{}_n)^m$. In particular, each $T^0{}_n$ contains exactly one mapping.† As a matter of convenience, we will use elements of $(T^1{}_n)^m$ as a way of referring to the corresponding elements of $T^m{}_n$.

† Lawvere's definition is equivalent; he prefers to work with the dual category; in this case, the coordinate morphisms can be identified by a product preserving functor from the dual of the category of natural numbers [LAVW63].

Exercise 3.14. In any algebraic theory, we have the following:

A. For each m, $n \geq 0$, let $\epsilon_{mn} = \langle \nu_{1\,n}, \ldots, \nu_{mn} \rangle \in T^m{}_n$. If $T^1{}_n \neq \emptyset$ (which is guaranteed for $n > 0$), let $\delta_{n+1\,n} = \langle \nu_{1\,n}, \ldots, \nu_{n\,n}, \xi \rangle$, where ξ is chosen arbitrarily from $T^1{}_n$; then $\epsilon_{n\,n+1}\delta_{n+1\,n}$ is the identity in $T^n{}_n$.

B. If $k \leq m \leq n$, then $\epsilon_{km} \circ \epsilon_{mn} = \epsilon_{kn}$.

C. If i, $j \leq n$, and $\theta_{ij} = \langle \nu_{1\,n}, \ldots, \nu_{i-1\,n}, \nu_{j\,n}, \nu_{i+1\,n}, \ldots, \nu_{n\,n} \rangle$, then $\nu_{in} \circ \theta_{ij} = \nu_{jn}$.

D. For $\chi \in T^m{}_n$ and $a > 0$, let $\chi^{+a} \in T^{m+a}{}_p$ be given by $p = \max(m + a, n)$, $\chi^{+a} = \langle \nu_{1\,m} \circ \chi \circ \epsilon_{np}, \ldots, \nu_{mm} \circ \chi \circ \epsilon_{np}, \nu_{m+1\,p}, \ldots, \nu_{m+a\,p} \rangle$. Then,
   $\epsilon_{m\,m+a} \circ \chi^{+a} = \chi \circ \epsilon_{np}$, so that χ is recoverable from χ^{+a} and m.
   $(\chi \circ \epsilon_{np})^{+a} = \chi^{+a}$.
   $(\chi \circ \epsilon_{np+1})^{+a} = \chi^{+a} \circ \epsilon_{p\,p+1}$.

Proposition 3.15. For any untyped instantiation system $\mathcal{A}$, $\mathcal{A}^C$ expands to an algebraic theory $\mathcal{A}^T$ that is unique up to isomorphism. Moreover, the mapping $\mathcal{A} \mapsto \mathcal{A}^T$ induces a bijection from isomorphism classes of untyped instantiation systems to isomorphism classes of algebraic theories.

Proof. To begin with, $\mathcal{A}^C$ really is a category, since composition is well defined: in the above definition, we have

$$\mathrm{var}(v\sigma\tau) \subseteq \bigcup\{\mathrm{var}(u\tau) \mid u \in \mathrm{var}(v\sigma)\} \subseteq \{\mathrm{var}(u\tau) \mid u \in V_p\} \subseteq V_n,$$

for each $v \in V_m$, as a result of Proposition 3.9A. For each i = 1, ..., m, let $\nu_{im} = \langle 1, v_i/v_1, m \rangle \in T^1{}_m$. If $\chi = \langle m, \sigma, n \rangle \in T^m{}_n$, then σ can be written uniquely in the form $\sigma = v_1\sigma/v_1 + \ldots + v_m\sigma/v_m$. This shows that the mapping $\chi \mapsto \langle \nu_{1m} \circ \chi, \ldots, \nu_{mm} \circ \chi \rangle$ is bijective.

We next define a mapping $\mathcal{T} \mapsto \mathcal{T}^I$ from algebraic theories to untyped instantiation systems. Let Q be the set of all maximal sets of the form $\{\psi, \psi \circ \epsilon_{n\,n+1}, \psi \circ \epsilon_{n\,n+2}, \ldots\}$, where $\psi \in T^m{}_n$. Notice that ψ is recoverable from any element in the set, by Exercise 3.14A (If $\psi \in T^m{}_0$, with $m > 0$, then $\nu_{1m} \circ \psi \in T^1{}_0$, so that δ_0 exists in this case). Each mapping $\chi \in T^m{}_p$ belongs to a unique element of Q: we can work back from p to obtain a maximal set of the form $\{ \ldots, \chi, \chi \circ \epsilon_{p\,p+1}, \ldots \}$; if χ is not the first element, its predecessor, χ', must be such that $\chi' \circ \epsilon_{p-1\,p} = \chi$, in which case, $\chi' = \chi \circ \delta_{p\,p-1}$. Let $\Phi(\chi)$ be the unique element of Q containing χ; Φ is one-to-one on each $T^m{}_p$, because no two elements of $\Phi(\chi)$ have the same codomain. The *terms* of $\mathcal{T}^I$ are those sets in Q of the form $\Phi(\chi)$, where $\chi \in T^1{}_p$, for some p.

For each $t \in Q$, let $t^+ = \{\chi^{+a} \mid \chi \in t, a \geq 0\}$; let $M = \{t^+ \mid t \in Q\}$; let S be the set of all maximal sets in M; S is essentially the set of all *substitutions* of $\mathcal{T}^I$. Each mapping $\chi \in T^n{}_p$ belongs to a unique element of S: one first constructs a maximal sequence of the form $\{ \ldots, \chi, \chi^{+1}, \chi^{+2}, \ldots \}$ with first element $\psi \in T^k{}_p$, for some k. Using Exercise 3.14D, we see that ψ is unique and that $\Phi(\psi)^+$ is a maximal element of M containing ψ and, hence, χ. Let $\Psi(\chi)$ be the unique element of S containing χ; $\Psi(\chi)$ always contains a unique element with minimal domain and codomain, which we refer to as the *minimal representative* of χ.

We can show that Ψ is one-to-one on each $T^p{}_q$: pick χ, $\chi' \in T^p{}_q$, and assume $\Psi(\chi) = \Psi(\chi')$. Let $\psi \in T^m{}_n$ be the common minimal representative of χ and χ'. Then $\chi = (\psi \circ \epsilon_{n\,n+i})^{+a}$, and $\chi' = (\psi \circ \epsilon_{n\,n+j})^{+b}$, for some a, b, i, and j. We have $p = m + a = m + b$, so that $a = b$. If $i \neq j$, say $i < j$, then $q = \max(m + a, n + i) = \max(m + a, n + j)$, so that $q = m + a \geq n + j$. Hence, $\chi' = (\psi \circ \epsilon_{n\,n+i} \circ \epsilon_{n+i\,n+j})^{+a} = (\psi \circ \epsilon_{n\,n+i})^{+a} = \chi$, by Exercises 3.14B and 3.14D.

Instantiation is defined by the formula $\Phi(\psi)\Psi(\chi) = \Phi(\psi' \circ \chi')$, where $\psi' \in \Phi(\psi)$, $\chi' \in \Psi(\chi)$, $\psi' \in T^1{}_p$, and $\chi' \in T^p{}_q$, for some ψ', χ', p, and q. Instantiation is well defined as a result of Exercise 3.14D. Composition of substitutions is defined similarly, and the associative laws follow directly from associativity for composition of mappings. The term $\Phi(\epsilon_0)$ is a monoid identity, and the set $V = \{\Phi(\nu_{ii}) \mid i > 0\}$ is easily seen to be a discriminating set for this monoid action, which establishes faithfulness. The *variables* of V are all equivalent by Exercise 3.14C, and every term is an instance of every variable. Pick $\chi \in T^m{}_n$; pick $i > m$. Then $\Phi(\nu_{ii})\Psi(\chi) = \Phi(\nu_{ii})$, because of the way in which χ^{+a} was defined. Consequently, if the constructed monoid action is represented as an action by substitutions on V, then $\mathrm{dom}(\Psi(\chi)) \subseteq V_m = \{\Phi(\nu_{11}), \ldots, \Phi(\nu_{mm})\}$. Moreover, if $k \leq m$, then $\Psi(\chi) | V_k = \Psi(\langle \nu_{1m} \circ \chi, \ldots, \nu_{km} \circ \chi \rangle)$. Hence, if $\psi \in T^1{}_k$, then V_k is a discriminating set for $\Phi(\psi)$.

To see that $(\mathcal{A}^T)^I$ is isomorphic to $\mathcal{A}$, we note that $\langle 1, t/v_1, m \rangle$ is recoverable from m and $\Phi(\langle 1, t/v_1, m \rangle)$, and that $\langle m, \sigma, n \rangle$ is recoverable from m, n, and $\Psi(\langle m, \sigma, n \rangle)$. Moreover, $\Phi(\langle 1, t/v_1, m \rangle)\,\Psi(\langle m, \sigma, n \rangle) = \Phi(\langle 1, t\sigma/v_1, n \rangle)$, and $\Phi(\langle 1, v_i/v_1, m \rangle) = \Phi(\langle 1, v_i/v_1, i \rangle) = \Phi(\nu_{ii})$, so that $\sigma \mapsto \Psi(\langle \mathrm{dom}(\sigma), \sigma, \mathrm{cdm}(\sigma) \rangle)$ becomes an isomorphism of instantiation systems when substitutions in $(\mathcal{A}^T)^I$ are represented as minimal finite functions on V.

It remains to show that $\mathcal{T} \mapsto (\mathcal{T}^I)^T$ is always isomorphic to $\mathcal{T}$. The requisite isomorphism is given, for each $T^m{}_n$, as $\chi \mapsto \langle m, \Psi(\chi), n \rangle$; we know it is well-defined, because $\mathrm{dom}(\Psi(\chi)) \subseteq V_m$ and V_n discriminates for $\Psi(\nu_{im} \circ \chi)$, for each $i \leq m$. It is one-to-one, because χ is recoverable from m, n, and $\Psi(\chi)$. To see that it is onto, it suffices to pick any mapping $\langle p, \Psi(\chi), q \rangle$ in $(\mathcal{T}^I)^T$ and show that $\Psi(\chi)$ contains a mapping from p to q. Let $\psi \in T^m{}_n$ be the minimal representative for χ. We know $\mathrm{dom}(\Psi(\psi)) \subseteq V_m$; if we had $\mathrm{dom}(\Psi(\psi)) \subseteq V_{m-1}$, we would have $\Phi(\nu_{mm})\Psi(\chi) = \Phi(\nu_{mm})$, implying that $\nu_{mm} \circ \psi = \nu_{mn}$, so that $\psi = \langle \nu_{1m} \circ \psi, \ldots, \nu_{m-1\,m} \circ \psi, \nu_{mn} \rangle = \langle \nu_{1m} \circ \psi, \ldots, \nu_{m-1\,m} \circ \psi \rangle^{+1}$, contradicting minimality of m. Hence, $p \geq m$. We know V_n is a discriminating set for each $\Phi(\nu_{im} \circ \psi)$; if V_{n-1} were also a discriminating set, we would have $\psi \circ \delta_{n\,n-1} \circ \epsilon_{n-1\,n} = \psi$, contradicting the assumption that ψ is the minimal representative for χ. We conclude that if $p = m$, then $q \geq n$, and if $p > m$, then $q \geq \max(n, p)$, so that $\Psi(\psi)$ contains a morphism from p to q. □

SECTION 4
CLASSIFICATION PROPERTIES

This section identifies properties associated with variable dependencies, type strictness, and well-foundedness of terms. These properties lead directly to a corresponding classification of instantiation systems and provide a basis for discussing fixed points of substitutions. Many of the presented results were previously obtained by Schmidt–Schauss and Siekmann for untyped systems [SCHM88, Sec. 9.1].

VARIABLE DEPENDENCY PROPERTIES

There are two classes of systems where the var operator is particularly well behaved. In systems where substitutions are *variable-preserving*, one has a useful characterization of var(tσ). These systems properly include those where terms *have unique quotients*.

A substitution σ *preserves variable occurrences* iff for any t, var(tσ) = ∪{var(uσ) | u ∈ var(t)}. We say substitutions are *variable-preserving* iff all substitutions preserve variable occurrences. Substitutions are variable-preserving in any order-sorted instantiation system. This is not true of lambda instantiation (Example 2.5), however: If t = [**f(y)**] and σ = [(**λ x . C**)/**f**], then tσ = [**C**], so that **y** ∈ ∪{var(uσ) | u ∈ var(t)} \ var(tσ).

In Proposition 4.1, Assertion A can fail, if η is merely a weakening for t (Example 5.1C). Assertion B can also fail, if the hypothesis is weakened. In general, renaming substitutions needn't preserve variable occurrences (Example 4.1A); invertible substitutions needn't either (Example 4.1B). Assertion D is also sharp (Example 4.1A again).

Proposition 4.1.

A. If η is a renaming for t (that is, a weakening for t that is a renaming), then var(tη) = {vη | v ∈ var(t)}.

B. Every invertible renaming substitution preserves variable occurrences.

C. Substitutions are variable-preserving iff for all t, var(t) = {v | for all σ, var(vσ) ⊆ var(tσ)}.

D. If σ is variable-preserving and var(s) ⊆ var(t), then var(sσ) ⊆ var(tσ).

Proof. Regarding Assertion A, we know var(tη) ⊆ {vη | v ∈ var(t)}, by Proposition 3.9A. We also know $(\eta\eta^{\vee})|\mathrm{var}(t) = (\eta^{\vee}\eta)|\mathrm{var}(t\eta) = \epsilon$, by Exercise 3.4B. We have,

$$\begin{aligned}
\mathrm{var}(t\eta) &\subseteq \{v\eta \mid v \in \mathrm{var}(t)\} \\
&= \{v\eta \mid v \in \mathrm{var}(t\eta\eta^{v})\} \\
&\subseteq \{v\eta \mid v \in \{u\eta^{v} \mid u \in \mathrm{var}(t\eta)\}\} \\
&= \{u\eta^{v}\eta \mid u \in \mathrm{var}(t\eta)\} \\
&= \mathrm{var}(t\eta).
\end{aligned}$$

Assertion B follows directly from Assertion A and Exercise 3.4B. To prove Assertion C, first assume $\mathrm{var}(t) = \{v \mid \text{for all } \sigma, \mathrm{var}(v\sigma) \subseteq \mathrm{var}(t\sigma)\}$. Pick $u \in \mathrm{var}(t)$; pick $v \in \mathrm{var}(u\sigma)$. By assumption, $\mathrm{var}(u\sigma) \subseteq \mathrm{var}(t\sigma)$, and thus $v \in \mathrm{var}(t\sigma)$. Conversely, suppose $\mathrm{var}(t\sigma) = \bigcup\{\mathrm{var}(u\sigma) \mid u \in \mathrm{var}(t)\}$, for all t and σ. Then, for each $u \in \mathrm{var}(t)$, $\mathrm{var}(u\sigma) \subseteq \mathrm{var}(t\sigma)$. Hence, $\mathrm{var}(t) = \{v \mid \text{for all } \sigma, \mathrm{var}(v\sigma) \subseteq \mathrm{var}(t\sigma)\}$. Assertion D is proved similarly. □

We can extend the definition of / as follows: Let t/s be the unique substitution σ such that $s\sigma = t$ and $\mathrm{dom}(\sigma) \subseteq \mathrm{var}(s)$, provided this is defined. We say that a term t *has unique quotients* iff t/s is defined, whenever $s \leq t$. Instantiation systems where all terms have unique quotients include first-order terms, tree instantiation systems, order-sorted instantiation systems, and Examples 4.2A and 4.2B below.

A substitution σ is *nearly invertible* iff there is a substitution τ such that $\mathrm{dom}(\tau) = \mathrm{cdm}(\sigma)$, $\mathrm{dom}(\sigma) = \mathrm{cdm}(\tau)$, and $(\sigma\tau)|\mathrm{dom}(\sigma) = (\tau\sigma)|\mathrm{dom}(\tau) = \epsilon$. Invertible substitutions and renaming substitutions are both nearly invertible in this sense.

Sharpness for Proposition 4.2A is illustrated by Example 2.3 (string instantiation): Let σ, τ, and t be given by σ = 'u**c**'/u, τ = '**c**v'/v, t = 'uv'. Then $t\sigma = t\tau$ = 'u**c**v'; the third set in Proposition 4.2A is empty, whereas the second is {u, v}. The nearly invertible substitution given in Proposition 4.2C needn't be a renaming substitution (see Example 4.1A). Moreover, the restriction to systems where terms have unique quotients cannot be dropped (see Example 4.3).

Proposition 4.2.

A. $\mathrm{var}(t) \supseteq \{v \mid \text{for all } \sigma, \mathrm{var}(v\sigma) \subseteq \mathrm{var}(t\sigma)\}$
$\supseteq \{v \mid \text{for all } \sigma, \tau, \text{ if } v\sigma \neq v\tau, \text{ then } t\sigma \neq t\tau\}$.

B. Terms have unique quotients iff all three sets in Assertion A are equal. In this case, substitutions are variable-preserving, by Proposition 4.1C.

C. If terms have unique quotients, $s \leq t$ and $t \leq s$, then $s = t\rho$, for some nearly invertible substitution ρ.

Proof. The first inclusion of Assertion A is obvious. To prove the second, pick $v \in \{v \mid \text{for all } \sigma, \tau, \text{ if } v\sigma \neq v\tau, \text{ then } t\sigma \neq t\tau\}$; pick σ; if possible, pick $u \in \mathrm{var}(v\sigma)$, and choose s so that $(v\sigma)(s/u) \neq v\sigma$. Then, by assumption, $t\sigma(s/u) \neq t\sigma$, showing that $u \in \mathrm{var}(t\sigma)$. Hence, $v \in \{v \mid \text{for all } \sigma, \mathrm{var}(v\sigma) \subseteq \mathrm{var}(t\sigma)\}$. For Assertion B, first assume terms have unique quotients. Pick $v \in \mathrm{var}(t)$. Suppose $t\sigma = t\tau$. Then both $\sigma|\mathrm{var}(t)$

and $\tau|\mathrm{var}(t)$ satisfy the definition of $(t\sigma)/t$. Consequently, $\sigma|\mathrm{var}(t) = \tau|\mathrm{var}(t)$. In particular, $\sigma|\{v\} = \tau|\{v\}$; that is, $v\sigma = v\tau$. We have just shown that $v \in \{v \mid \text{for all } \sigma, \tau, \text{ if } v\sigma \neq v\tau, \text{ then } t\sigma \neq t\tau\}$. Conversely, assume $\mathrm{var}(t) = \{v \mid \text{for all } \sigma, \tau, \text{ if } v\sigma \neq v\tau, \text{ then } t\sigma \neq t\tau\}$, for all terms t. Suppose $s \leq t$. Let σ, τ be any two substitutions such that $s\sigma = s\tau = t$. Then, by assumption, $v\sigma = v\tau$, for all $v \in \mathrm{var}(s)$. That is, $\sigma|\mathrm{var}(s) = \tau|\mathrm{var}(s) = t/s$, so we have unique quotients.

For Assertion C, use Lemma 3.5 to choose a renaming substitution η such that $\mathrm{dom}(\eta) = \mathrm{var}(t)$ and $\mathrm{var}(s)$ is disjoint from $\mathrm{var}(t\eta)$. Then $t\eta \leq t\eta\eta^{\vee} = t \leq s$. Let $\pi = (t\eta/s + s/t\eta)$. Clearly, $s\pi = t\eta$ and $t\eta\pi = s$. To see that π is invertible, in fact self-invertible, pick $v \in \mathrm{dom}(\pi)$. If $v \in \mathrm{var}(s)$, for example, then $\mathrm{var}(v) \subseteq \mathrm{var}(s)$ and, by Proposition 4.1D, $\mathrm{var}(v(t\eta/s)) \subseteq \mathrm{var}(s(t\eta/s)) = \mathrm{var}(t\eta)$. Consequently,

$$\begin{aligned} v\pi^2 &= v(t\eta/s)\pi && \text{— } v \in \mathrm{var}(s). \\ &= v(t\eta/s)(s/t\eta) && \text{— } \mathrm{var}(v(t\eta/s)) \subseteq \mathrm{var}(t\eta). \\ &= v\epsilon = v && \text{— } s(t\eta/s)(s/t\eta) = s\epsilon \text{ and Proposition 4.2A} \end{aligned}$$

Thus, the required nearly invertible substitution is $\rho = \eta\pi$. □

TYPE STRICTNESS PROPERTIES

Systems in which terms are untyped or are *strictly typed* are of historical interest. Systems where terms are *restrictable* or are *weakly restrictable* have better unification algorithms. A form of *weak restrictability* is quite feasible in languages where nonempty type-intersection is one of the basic type constructors. Comparison of these various *type strictness* properties is simplified, if we consider only those systems where every term is an instance of a variable.

Define the *instance-type* of a variable v to be the set $T_v = \{t \mid v \leq t\}$. We say that

variables are untyped iff $T_u = T_v$ (for all u and v).
variables are strictly typed iff $T_u = T_v$ or $T_u \cap T_v = \emptyset$.
terms are restrictable iff $t\sigma \in T_v$ implies that, for some η such that η is a weakening for t, $t\eta \in T_v$, and $v\eta \leq v\sigma$, for all $v \in \mathrm{var}(t)$.
terms are weakly restrictable iff $t \in T_u \cap T_v$ implies $t \in T_w \subseteq T_u \cap T_v$, for some w.
terms are first-class iff $t \in T_v$, for some v.

The last three of these definitions make sense for particular terms as well as for instantiation systems. In any system, all terms are weakly restrictable iff all variables are restrictable. For those order-sorted systems where every type is an instance-type, restrictability is equivalent to weak restrictability (as a result of Exercise 3.12C).

Proposition 4.3. The above five properties are related as follows, where examples illustrating sharpness are given in parentheses:

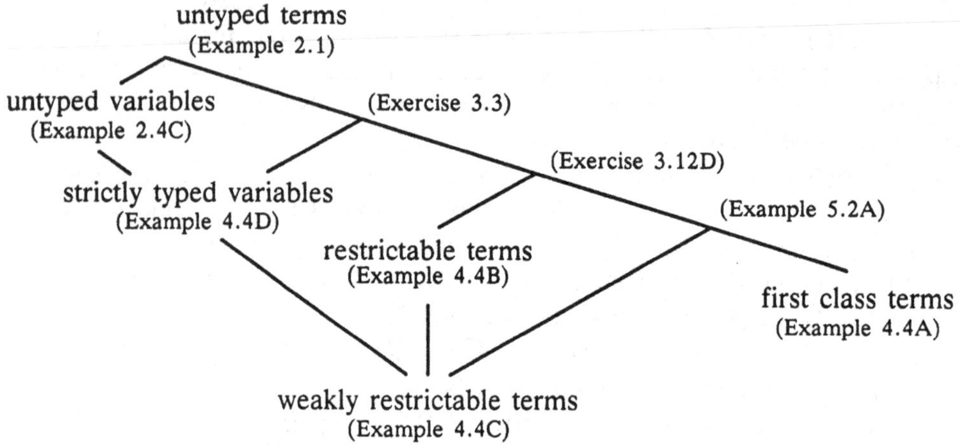

Proof. We prove only two of the above implications, as the others follow immediately from the definitions involved. To show that if variables are strictly typed and first-class, then terms are restrictable, assume $t\sigma \in T_v$. Choose u so that $t \in T_u$. Then $t\sigma \in T_u \cap T_v$, so $T_u = T_v$, so $t = t\epsilon \in T_v$, and $\epsilon\sigma = \sigma$. To show that if terms are restrictable, then they are weakly restrictable, assume $t \in T_u \cap T_v$. Since $u(t/u) \in T_v$, we may choose a weakening η so that $u\eta \in T_v$ and $u\eta \leq u(t/u) = t$. Taking $w = u\eta$, we have $T_w \subseteq T_u$, since $u \leq w$, and $T_w \subseteq T_v$, since $w = u\eta \in T_v$. Moreover, $t \in T_w$, since $w \leq t$. $\square$

TERM OCCURRENCE PROPERTIES

The following definition of *term-occurrence* generalizes that found in tree systems. In general, var(t) need not contain all variables that may occur in t, just those that t depends on. Terms may be classified according to whether subterm occurrences are *cyclic, acyclic, or well-founded.* Well-founded terms are acyclic and admit a convenient notion of *term rank.* Acyclic terms needn't be well-founded (see Example 4.5C below).

We say s *occurs in* t iff, for some σ, r, and $v \in \mathrm{var}(r)$, $t = r\sigma$ and $s = v\sigma$. In other words, s occurs in t iff t is of the form $t = r\langle \ldots s \ldots \rangle$, for some r. We say s *occurs properly in* t iff, in addition, $r \neq v$. A term occurs in itself iff it is an instance of a variable. In TR#(A, V), for example, s occurs in t iff s is a subtree of t.

The hypotheses of Proposition 4.4 cannot be dropped (see Examples 4.1A, 1C, and 1D and 4D below). Proposition 4.4B does not extend to proper occurrences (Example 4.5D below).

Proposition 4.4. Assume substitutions are variable-preserving.

A. v occurs in t iff $v \in var(t)$.

B. s occurs in t iff $t = r(s/v)$, for some r and v, with $v \in var(r)$.†

C. If terms are restrictable, then the occurs-in relation is transitive.

Proof. For Assertion A, if $v \in var(t)$, then $t = t\epsilon$, with $v \in var(t)$. Conversely, if $t = r\sigma$, $v = u\sigma$, and $u \in var(r)$, then $var(u\sigma) \subseteq var(r\sigma) = var(t)$, by Proposition 4.1D. For Assertion B, if $t = r(s/v)$, with $v \in var(r)$, then s clearly occurs in t. Conversely, if s occurs in t, then we may choose σ, r, and v so that $t = r\sigma$, $s = v\sigma$, $v \in var(r)$, and $v \notin cdm(\sigma)$. Let $r' = r(\sigma|(V \setminus \{v\}))$. Then $t = r\sigma = r(\sigma|(V \setminus \{v\}))(s/v) = r'(s/v)$. Finally, $\{v\} = var(v(\sigma|(V \setminus \{v\})) \subseteq var(r')$, because substitutions are variable-preserving.

For Assertion C, suppose r occurs in s and s occurs in t. Using Assertion B, we may choose s′, t′, u, v, so that $s = s'(r/u)$, $u \in var(s')$, $t = t'(s/v)$, and $v \in var(t')$. Since $v \leq s = s'(r/u)$, we may choose a weakening w/u such that $v \leq s'(w/u)$, $w \leq r$, and $w \notin var(s') \cup var(t')$. We see easily that $((w/u)(r/w))|var(s') = r/u$, and that $((s'(w/u)/v)(r/w))|var(t') = s'(r/u)/v = s/v$. Consequently, $t'(s'(w/u)/v)(r/w) = t'(s/v) = t$. Finally, using the fact that substitutions are variable-preserving, we see that $w \in var(s'(w/u))$, since $u \in var(s')$, and that $w \in var(t'(s'(w/u)/v))$, since $v \in var(t')$. We have thus shown that r occurs in t. □

A term t_1 is *(occurrence) well-founded* iff there does not exist an infinite sequence $t_1, t_2, \ldots$ such that $t = t_1$, and each t_{i+1} properly occurs in t_i; it is *acyclic* iff there does not exist a sequence $t_1, \ldots, t_n$ such that t_2 properly occurs in t_1; ... t_n properly occurs in t_{n-1}; t_1 properly occurs in t_n; and each t_i occurs in t. The *rank* of a well-founded term t is recursively defined to be the least ordinal greater than the rank of any term that properly occurs in t. The terms of order-sorted systems are well-founded; the terms of rational tree systems may be cyclic. The terms of Example 2.4C are well-founded, whereas every term of Example 2.5 occurs properly in itself: $[t] = [\mathbf{f(C)}][(\lambda\, \mathbf{x} \,.\, t)/\mathbf{f}]$, provided **x** is not free in t. Well-foundedness is compared with several similar properties in Examples 4.2A and 4.5 below. Sharpness of Proposition 4.5A is illustrated by Example 4.5E.

Proposition 4.5.

A. If t and $t\sigma$ are well-founded, then $rank(t) \leq rank(t\sigma)$.

B. If terms are acyclic, then no variable is an instance of a nonvariable [cf BURC87, Sec. 3].

Proof: For assertion A, suppose $rank(t) = \alpha$. If $\alpha = 0$, then, trivially, $0 \leq rank(t\sigma)$. Otherwise, Pick $\beta < \alpha$. We may choose r, τ, and $v \neq r$ so that $t = r\tau$, $v \in var(r)$, and $\beta \leq rank(v\tau) < \alpha$. We have,

† This gives the relationship between the occurs-in relation and the similar but more restrictive subterm relation of [SCHM88, Definition 9.1.2].

$$\beta \leq \text{rank}(v\tau) \leq \text{rank}(v\tau\sigma) < \text{rank}(r\tau\sigma) = \text{rank}(t\sigma),$$

using choice of v and τ, ordinal induction, $v\tau\sigma$ properly occurs in $r\tau\sigma$, and $r\tau\sigma = t\sigma$. Hence $\text{rank}(t) = \alpha \leq \text{rank}(t\sigma)$.

Suppose Assertion B were false. Choose t, σ, and v so that $t \notin V$, $t\sigma = v$, and dom(σ) is as small as possible. In particular, $\text{dom}(\sigma) \subseteq \text{var}(t)$. We know $\{v\} = \text{var}(t\sigma) \subseteq \bigcup\{\text{var}(u\sigma) \mid u \in \text{var}(t)\}$, by Proposition 3.9A. Choose $u \in \text{var}(t)$ so that $v \in \text{var}(u\sigma)$. Then v occurs in $u\sigma$ and $u\sigma$ occurs properly in $t\sigma = v$. If v occurs properly in $u\sigma$, we have an occurrence cycle of length 2. If v occurs improperly in $u\sigma$, then $v = u\sigma$, so that v occurs properly in itself, and we have an occurrence cycle of length 1. □

ITERATION AND FIXED POINTS

The following results are not essential to later sections; they provide a concise theory that mirrors certain problems encountered in the design of unification algorithms. As such, they provide counterexamples illustrating sharpness of unification results and suggest useful alternatives to the presented algorithms.

Interest in fixed points stems from the fact that if τ is a fixed point of σ, then τ unifies t and $t\sigma$, for any t. Often, most-general fixed points of σ take the form of *iterates*. At the end of Section 2, we saw how iteration could be used to abstractly model graph implementations of tree systems. In the case of finite tree systems, the iterate σ^* can be understood in terms of finite limits, as being σ^n, for sufficiently large n. For rational tree systems, σ^* still coincides with σ^∞ in some obvious sense, but intuition is strained by the the fact that σ^* is computable in linear time. Finally, there are other systems where the limit approach seems to break down completely, and it is necessary to use a general *fixed-point* approach similar to that introduced by Elgot [ELGO75]. Fortunately, this approach is also the most elegant.

Of the following results, Theorem 4.6 relates iteration to the *occurs check* found in many unification algorithms. Sharpness for theorem 4.6 is partially supported by Examples 4.6A and 5.3A. Portions of Exercise 4.7 below were assumed in Elgot's definition of iteration. But, despite differences in formalization and content, the broad outlines of the main theorem by Bloom, Ginali, and Rutledge [BLOO77] can be found in Theorem 4.10. The proof of Theorem 4.10A uses the "brute force" method for solving simultaneous equations from high school algebra: to find a solution τ to the system $\tau = \sigma'\tau$, solve the first equation, $v\tau = v\sigma'\tau$, obtaining a term t such that $t = s(t/v)$, where $s = v\sigma'$. Substitute t for v in the remaining terms, and solve the resulting smaller system of the form $\kappa = \eta\kappa$. Finally, plug these solutions back into t. Sharpness for Theorem 4.10C is supported by Example 4.6B.

A *fixed-point* of σ is a substitution τ such that $\sigma\tau = \tau$. Notice that if u and v are equivalent, then v/u is a fixed point of u/v, by Exercise 3.2C. A fixed point of σ whose domain is a subset of σ is a *proper* fixed point. We write σ^* for the unique proper fixed point of σ, provided there is exactly one proper fixed point; this is the *iterate* of σ. In a rational tree system, the substitution **(1 + x)/x** iterates to **(1 + (1 + (...)))/x**, where **1 + (1 + (...))** is regarded as an infinite rational tree. In Example 2.5, the substitution **[(λ u.u(x(λv.v)))/x]** has infinitely many proper fixed points, including **[(λ u.u(y))/x]** [HUET75].

A substitution σ *passes the occurs check* iff there do not exist $u_1, \ldots, u_k \in \mathrm{dom}(\sigma)$ such that $u_1 \in \mathrm{var}(u_2\sigma), \ldots, u_{k-1} \in \mathrm{var}(u_k\sigma)$, and $u_k \in \mathrm{var}(u_1\sigma)$. A *triangular enumeration* of σ is a sequence $\sigma_1, \ldots, \sigma_n$ such that $\sigma = \sigma_1 + \ldots + \sigma_n$ and $\mathrm{dom}(\sigma_i) \cap \mathrm{cdm}(\sigma_j) = \emptyset$, whenever $i \leq j$. If $\sigma_1, \ldots, \sigma_n$ is a triangular enumeration of σ, then $\sigma^* = \sigma_1 \ldots \sigma_n$.†

We say σ is *permutation-free* iff the variable-valued portion of σ passes the occurs check, that is, iff there do not exist $u_1, \ldots, u_k \in \mathrm{dom}(\sigma)$ such that $u_1 = u_2\sigma, \ldots, u_{k-1} = u_k\sigma$, and $u_k = u_1\sigma$. In other words, σ is permutation-free iff ϵ is the only permutation of the form $\sigma|X$.

Theorem 4.6. Of the following conditions, A and B are equivalent. They imply conditions C and D. If substitutions are variable-preserving, then Conditions A through D are equivalent. Conditions C and D together imply Condition E, which implies Condition F. If terms are acyclic, then all six conditions are equivalent (even if substitutions aren't variable preserving).

A. σ passes the occurs check.
B. σ has a triangular enumeration.
C. $v \notin \mathrm{var}(v\sigma^k)$, for all $v \in \mathrm{dom}(\sigma)$ and all $k > 0$.
D. $\mathrm{cdm}(\sigma^n) \cap \mathrm{dom}(\sigma) = \emptyset$, whenever $n \geq \mathrm{size}(\mathrm{dom}(\sigma))$.
E. $\sigma^* = \sigma^n$, where $n = \mathrm{size}(\mathrm{dom}(\sigma))$.
F. σ is permutation-free, and $\sigma\tau = \tau$, for some τ.

Proof. The equivalence of conditions A and B is just a general fact about finite binary relations. To see that A implies B, for example, first notice that condition A implies there exists $v_n \in \mathrm{dom}(\sigma)$ such that $v \notin \mathrm{var}(v_n\sigma)$, for all $v \in \mathrm{dom}(\sigma)$. A triangular enumeration of the form $v_1\sigma/v_1, \ldots, v_n\sigma/v_n$ may be constructed by choosing $v_{n-1} \in \mathrm{dom}(\sigma)$ so that $v \notin \mathrm{var}(v_{n-1}\sigma)$, for all $v \in \mathrm{dom}(\sigma) \setminus \{v_n\}$, and so forth.

B implies C and D: Assume a triangular enumeration of the form $v_1\sigma/v_1, \ldots, v_n\sigma/v_n$. We have directly that $\mathrm{var}(v_j\sigma) \cap \mathrm{dom}(\sigma) \subseteq \{v_i \mid i \geq j + 1\}$. Assume, by induction, that $\mathrm{var}(v_j\sigma^k) \cap \mathrm{dom}(\sigma) \subseteq \{v_i \mid i \geq j + k\}$. We have,

† Moreover, σ^* can be computed in time proportional to the weight of the first n − 1 terms, assuming that compositions are evaluated from right to left and that repeated subterms need only be stored once, so that each variable replacement takes unit time.

$$\begin{aligned} var(v_j\sigma^{k+1}) \cap dom(\sigma) &\subseteq \bigcup \{var(w\sigma) \cap dom(\sigma) \mid w \in var(v_j\sigma^k) \cap dom(\sigma)\} \\ &\subseteq \bigcup \{var(v_i\sigma) \cap dom(\sigma) \mid i \geq j + k\} \\ &\subseteq \{v_m \mid m \geq j + k + 1\}. \end{aligned}$$

This gives us both conditions C and D as special cases.

C and variable-preservation imply A: Suppose A were false. Choose $u_1, ..., u_k \in dom(\sigma)$ so that $u_1 \in var(u_2\sigma), ..., u_{k-1} \in var(u_k\sigma)$, and $u_k \in var(u_1\sigma)$. Since substitutions are variable-preserving, we have, $u_1 \in var(u_2\sigma) \subseteq var(u_3\sigma^2) \subseteq ... \subseteq var(u_k\sigma^{k-1}) \subseteq var(u_1\sigma^k)$, by Proposition 4.1D. This contradicts condition C.

D and variable-preservation imply C: Choose n so that $cdm(\sigma^n) \cap dom(\sigma) = \emptyset$. Then for all m, $cdm(\sigma^{n+m}) \cap dom(\sigma) = \emptyset$. Suppose Condition C were false. Choose v, k so that $v \in dom(\sigma) \cap var(v\sigma^k)$. Since substitutions preserve variable occurrences, we have $var(v) \subseteq var(v\sigma^k) \subseteq var(v\sigma^{2k}) \subseteq ... \subseteq var(v\sigma^{nk})$, for all n, which gives the desired contradiction.

C and D imply E: First, $dom(\sigma^n) = dom(\sigma)$, by Assertion C. This and Assertion D imply that $\sigma\sigma^n = \sigma^n\sigma = \sigma^n$. If τ is also a proper fixed point, then $\tau = \sigma^n\tau$; moreover, $\sigma^n\tau = \sigma^n$, since $cdm(\sigma^n) \cap dom(\tau) = \emptyset$. Hence, $\tau = \sigma^n = \sigma^*$.

E implies F: To see that σ is permutation-free, suppose otherwise. Then, for some $v \in dom(\sigma)$ and $k > 0$, we have $v\sigma^k = v$. But then, $v\sigma^n = v\sigma^{nk} = v$, and $v\sigma^n = v\sigma\sigma^n = v\sigma$, contradicting $v \in dom(\sigma)$.

F and terms are acyclic imply A: Assume Condition F. Since σ is permutation-free, we may recursively define ζ as follows: $v\zeta = v\sigma$, if $v\zeta \notin dom(\sigma)$, and $v\zeta = v\sigma\zeta$, otherwise. Then $v\zeta \notin dom(\sigma)$, for all v, so that $dom(\zeta) = dom(\sigma)$. A trivial finite induction shows $v\zeta\tau = v\tau$, for all $v \in dom(\zeta)$. We will first show that ζ satisfies the occurs check. Suppose otherwise; choose $u_1, ..., u_k \in dom(\zeta)$ so that $u_1 \in var(u_2\zeta), ..., u_{k-1} \in var(u_k\zeta)$, and $u_k \in var(u_1\zeta)$. Then, $u_1\tau$ occurs in $u_2\tau$, since $u_2\tau = (u_2\zeta)\tau$, and $u_1 \in var(u_2\zeta)$; this occurrence is proper, since otherwise, we would have $u_1 = u_2\zeta \notin dom(\sigma)$. Similarly, $u_2\tau$ occurs properly in $u_3\tau$, ..., and $u_k\tau$ occurs properly in $u_1\tau$, so that $u_1\tau$, for example, is not acyclic. We now may choose a triangular enumeration of ζ of the form $v_1\zeta/v_1, ..., v_n\zeta/v_n$. If $i < j < n$, and $v_i\sigma = v_n$, then we may place $v_i\zeta/v_i$ after $v_j\zeta/v_j$ in the triangular enumeration, because $v_j \notin var(v_n\zeta) = var(v_i\zeta)$. Thus, we may assume a triangular enumeration of the form $v_1\zeta/v_1, ..., v_k\zeta/v_k, ..., v_n\zeta/v_n$, where $v_k\sigma = v_{k+1}, ..., v_{n-1}\sigma = v_n$. A trivial induction shows that further re-arrangements allow us to assume that if $v_i\sigma \in dom(\sigma)$, then $v_{i+1} = v_i\sigma$. At this point, $v_1\sigma/v_1, ..., v_n\sigma/v_n$ is a triangular enumeration of σ. Hence, σ passes the occurs check. □

Exercise 4.7. Let τ be a fixed point of σ that fails to satisfy one of the following conditions:

A. $\mathrm{dom}(\tau) \supseteq \mathrm{dom}(\sigma)$.

B. $\mathrm{cdm}(\tau) \cap \mathrm{dom}(\sigma) = \emptyset$.

C. $\mathrm{cdm}(\tau) \subseteq \mathrm{cdm}(\sigma)$.

Then there are multiple fixed points τ' that satisfy conditions A and B. If $\mathrm{dom}(\tau) \subseteq \mathrm{dom}(\sigma)$, then Condition A may be further strengthened to $\mathrm{dom}(\tau') = \mathrm{dom}(\sigma)$. (See Example 5.3B regarding sharpness.)

Exercise 4.8. Assume σ^* is defined. Then

A. $\sigma^* = \sigma^*\sigma = \sigma^*\sigma^*$.

B. $(\sigma^n)^* = \sigma^*$, if $(\sigma^n)^*$ is defined.

C. $\mathrm{dom}(\sigma^*) = \mathrm{dom}(\sigma)$.

D. $\mathrm{cdm}(\sigma^*) \subseteq \mathrm{cdm}(\sigma) \setminus \mathrm{dom}(\sigma)$.

E. $\sigma^* = \sigma$ iff $\mathrm{dom}(\sigma) \cap \mathrm{cdm}(\sigma) = \emptyset$.

F. If σ^*, τ^* and $(\sigma + \tau)^*$ are defined and $\mathrm{dom}(\sigma) \cap \mathrm{cdm}(\tau) = \emptyset$, then $(\sigma + \tau)^* = \sigma^*\tau^*$.

Lemma 4.9. Assume no variable is an instance of a nonvariable. Assume $\mathrm{dom}(\rho) \subseteq \mathrm{dom}(\sigma)$, $\mathrm{dom}(\eta) \cap \mathrm{dom}(\sigma) = \emptyset$, and $\mathrm{cdm}(\eta) \cap \mathrm{dom}(\sigma) = \emptyset$.

A. $\sigma\eta\rho = \eta\rho$ iff ρ is a proper fixed point of $(\sigma\eta)|\mathrm{dom}(\sigma)$.

B. If ρ is a fixed point of σ, then $(\rho\eta)|\mathrm{dom}(\sigma)$ is a proper fixed point of $(\sigma\eta)|\mathrm{dom}(\sigma)$.

C. $(\sigma^*\eta)|\mathrm{dom}(\sigma) = ((\sigma\eta)|\mathrm{dom}(\sigma))^*$, if σ^* and $((\sigma\eta)|\mathrm{dom}(\sigma))^*$ are defined.

D. $(s/v)^*\eta = ((s\eta)/v)^*$, if both are defined and $v \notin \mathrm{dom}(\eta) \cup \mathrm{cdm}(\eta)$.

Proof. First, if $v \in \mathrm{dom}(\sigma)$, then $v\sigma\eta \neq v$. Suppose otherwise. By assumption, $v\sigma$ must be a variable, because $v\sigma\eta$ is. But then $v\sigma \in \mathrm{dom}(\eta)$, since $v\sigma\eta \neq v\sigma$. Hence, $v = v\sigma\eta \in \mathrm{cdm}(\eta)$, contradicting $\mathrm{cdm}(\eta) \cap \mathrm{dom}(\sigma) = \emptyset$. Consequently, $\mathrm{dom}((\sigma\eta|\mathrm{dom}(\sigma)) = \mathrm{dom}(\sigma) \supseteq \mathrm{dom}(\rho)$.

For Assertion A, if $\sigma\eta\rho = \eta\rho$, then $(\sigma\eta)|\mathrm{dom}(\sigma)\rho = (\sigma\eta\rho)|\mathrm{dom}(\sigma) = (\eta\rho)|\mathrm{dom}(\sigma) = \rho$. Conversely, suppose $(\sigma\eta)|\mathrm{dom}(\sigma)\rho = \rho$. If $v \in \mathrm{dom}(\sigma)$, then $v\sigma\eta\rho = v(\sigma\eta)|\mathrm{dom}(\sigma)\rho = v\rho = v\eta\rho$. If $v \notin \mathrm{dom}(\sigma)$, then $v\sigma\eta\rho = v\eta\rho$, directly.

For Assertion B, we have, already, that $\mathrm{dom}((\rho\eta)|\mathrm{dom}(\sigma)) \subseteq \mathrm{dom}(\sigma) = \mathrm{dom}((\sigma\eta)|\mathrm{dom}(\sigma))$. We need to show, for each $v \in \mathrm{dom}(\sigma)$, that $v\sigma\eta(\rho\eta)|\mathrm{dom}(\sigma) = v\rho\eta$. By assumption, $v\rho = v\sigma\rho$, so we need only show that $\eta(\rho\eta)|\mathrm{dom}(\sigma) = \rho\eta$. Pick u. If $u \in \mathrm{dom}(\sigma)$, then $u\eta(\rho\eta)|\mathrm{dom}(\sigma) = u\rho\eta$, directly. If $u \notin \mathrm{dom}(\sigma)$, then $\mathrm{var}(u\eta)$ is disjoint from $\mathrm{dom}(\sigma)$, so that $u\eta(\rho\eta)|\mathrm{dom}(\sigma) = u\eta = u\rho\eta$.

Assertion C follows directly from Assertion B, since $(\sigma^*\eta)|\mathrm{dom}(\sigma)$ is now the unique fixed point of $(\sigma\eta)|\mathrm{dom}(\sigma)$ whose domain is a subset of $\mathrm{dom}(\sigma)$. Finally, Assertion D follows directly from Assertion C, with $\sigma = s/v$. □

Theorem 4.10.

A. If scalar substitutions have proper fixed points, then every substitution has proper fixed points.

B. If scalar substitutions have proper fixed points, then every substitution that contains a permutation has infinitely many proper fixed points.

C. If no variable is an instance of a nonvariable, and scalar substitutions have iterates, then σ^* is defined iff σ is permutation-free.

Proof. Assertion A is proved by induction on the size of the domain of an arbitrary substitution σ. The case $\sigma = s/v$ follows directly from the definitions involved. Let $\sigma' = \sigma + s/v$, where $s \neq v$. Let t/v be a (proper) fixed point of s/v; let $\eta = (\sigma(t/v))|\text{dom}(\sigma)$. By induction, we may let κ proper be a proper fixed point of η. Let $\tau = (t/v)\kappa$. It remains to show that τ is a proper fixed point of σ'. Clearly, $\text{dom}(\tau) \subseteq \{v\} \cup \text{dom}(\kappa) \subseteq \text{dom}(\sigma')$. We prove $\tau = \sigma'\tau$ as follows:

$v\sigma'\tau = s(t/v)\kappa$	— definition of σ', τ
$= t\kappa$	— t/v is a fixed point of s/v
$= v(t/v)\kappa = v\tau$	— definition of τ.

If $u \in \text{dom}(\sigma)$, then

$u\sigma'\tau = u\sigma(t/v)\kappa$	— definition of σ', τ
$= u(\sigma(t/v))|\text{dom}(\sigma)\kappa$	— $u \in \text{dom}(\sigma)$
$= u\eta\kappa = u\kappa$	— definition of η, κ is a fixed point of η
$= u(t/v)\kappa = u\tau$	— $u \neq v$, definition of τ.

For Assertion B, suppose σ contains a permutation, say $\sigma|A$, where $A = \{v_1, \ldots, v_n\}$, $v_n = v$ and, for each i, $v_i = v\sigma^i$. By induction and Assertion A, we may let κ be a proper fixed point of the substitution $\sigma|(V \setminus A)$. Let z be a variable equivalent to v that does not belong to $\text{dom}(\sigma) \cup \text{cdm}(\sigma) \cup \text{cdm}(\kappa)$; let $\zeta = z/v_1 + \ldots + z/v_n$. Then $\text{dom}(\zeta) = A$, and $\text{dom}(\kappa\zeta) \subseteq \text{dom}(\sigma)$. Moreover, $\kappa\zeta = \sigma\kappa\zeta$: If $v \in A$, then $v\kappa\zeta = v\zeta = z = (v\sigma)\zeta = v\sigma\kappa\zeta$. If $v \notin A$, then $v\kappa\zeta = v\sigma|(V \setminus A)\kappa\zeta = v\sigma\kappa\zeta$. Finally, there are infinitely many fixed points, as can be seen by varying z.

For Assertion C, we show by induction, that any permutation-free substitution can only have one proper fixed point. The case $\sigma = s/v$ follows directly from the definitions. Let $\sigma' = \sigma + s/v$, where $s \neq v$ and σ' is permutation-free. Let τ' be a fixed point of σ'. We will show that τ' is the substitution constructed in the proof of Assertion A. More precisely, $\tau' = (s/v)^*\eta^*$, where $\eta = (\sigma(s/v)^*)|\text{dom}(\sigma)$. By Exercise 4.7B, we may prove this assuming τ' satisfies the additional condition, $\text{cdm}(\tau') \cap \text{dom}(\sigma') = \emptyset$.

1. $(\tau'|\text{dom}(\sigma))(v\tau'/v) = \tau'|\text{dom}(\sigma) + (v\tau')/v = \tau'$, using Corollary 3.8D, $v \notin \text{dom}(\sigma)$, and $v \notin \text{cdm}(\tau')$.
2. $v\tau' = v\sigma'\tau' = s\tau' = s(\tau'|\text{dom}(\sigma))(v\tau'/v)$, by choice of τ', choice of σ', and Step 1.

3. $(v\tau'/v) = (s\tau'|\mathrm{dom}(\sigma)/v)^*$: From Step 2 and the assumption that scalar fixed points are unique, it suffices to show that $s(\tau'|\mathrm{dom}(\sigma)) \neq v$. Suppose not. By assumption, $v \notin \mathrm{cdm}(\tau')$, so if $s(\tau'|\mathrm{dom}(\sigma)) = v$, then $v \in \mathrm{var}(s)$ by Proposition 3.9B. Since $v \neq s$, s is not a variable. But then $s(\tau'|\mathrm{dom}(\sigma))$ cannot be a variable, since no variable is an instance of a nonvariable. This gives the desired contradiction.
4. $\tau' = (s/v)^*(\tau'|\mathrm{dom}(\sigma))$, since

$$
\begin{aligned}
v(s/v)^*(\tau'|\mathrm{dom}(\sigma)) &= v(s\tau'|\mathrm{dom}(\sigma)/v)^* && \text{— Lemma 4.9D}\\
&= v\tau' && \text{— Step 3.}
\end{aligned}
$$

5. $\mathrm{dom}(\sigma) = \mathrm{dom}(\eta)$: If $u \in \mathrm{dom}(\sigma) \setminus \mathrm{dom}(\eta)$, then $u = u\eta = u\sigma(s/v)^*$ is a variable. Since no variable is an instance of a nonvariable, $u\sigma$ must be a variable too. This can happen only if $u\sigma = v$. But then $\sigma'|\{u, v\}$ is a permutation, contrary to assumption.
6. η is permutation-free: Suppose, to the contrary, that η contains a permutation, say $\eta|A$, where $A = \{v_1, ..., v_n\}$, $v_n = v_0$ and, for each i, $v_i = v_0\eta^i$. Then, for each i, $v_i = v_{i-1}\sigma(s/v)^*$, where $v_{i-1}\sigma$ must be a variable, since no variable is an instance of a nonvariable. Moreover, either $v_{i-1}\sigma = v$ and $v(s/v)^* = v_i$, or else $v_{i-1}\sigma = v_i$. Consequently, $\sigma'|\{v, v_1, ..., v_n\}$ contains a permutation, contrary to assumption.
7. $\tau'|\mathrm{dom}(\sigma) = \eta^*$: We know $\mathrm{dom}(\tau'|\mathrm{dom}(\sigma)) \subseteq \mathrm{dom}(\eta)$ by Step 5. By induction and Step 6, it remains to show that $\tau'|\mathrm{dom}(\sigma) = \eta(\tau'|\mathrm{dom}(\sigma))$. Pick $u \in \mathrm{dom}(\sigma) = \mathrm{dom}(\eta)$. Then

$$
\begin{aligned}
u\tau' &= u\sigma'\tau' && \text{— choice of } \tau'\\
&= u(\sigma + s/v)\tau' && \text{— choice of } \sigma'\\
&= u((\sigma\tau')|\mathrm{dom}(\sigma) + (s\tau')/v) && \text{— Exercise 3.1I}\\
&= u(\sigma\tau') && \text{— } u \in \mathrm{dom}(\sigma)\\
&= u\sigma(s/v)^*(\tau'|\mathrm{dom}(\sigma)) && \text{— Step 4}\\
&= u\eta(\tau'|\mathrm{dom}(\sigma)) && \text{— definition of } \eta.
\end{aligned}
$$

Hence, $\tau' = (s/v)^*\eta^*$, by Steps 4 and 7. □

COUNTEREXAMPLES

4.1. *Variable Dependencies and Variable Occurrences.*

A. Beginning with first-order term instantiation, take the quotient system induced by the equation $\mathbf{x} - \mathbf{x} = \mathbf{0}$. Let t be the term $[\mathbf{x} - \mathbf{y}]$, so that $\mathrm{var}(t) = \{\mathbf{x}, \mathbf{y}\}$. Let $\eta = \mathbf{x}/\mathbf{y}$. Regarding Proposition 4.1B, we have $\mathrm{var}(t\eta) = \mathrm{var}([\mathbf{0}]) = \emptyset$, while $\bigcup\{u\eta \mid u \in \mathrm{var}(t)\} = \{\mathbf{x}\}$, even though η is a renaming. Here, $\mathrm{var}(t) \supset \{v \mid \text{for all } \sigma, \mathrm{var}(v\sigma) \subseteq \mathrm{var}(t\sigma)\} = \emptyset$. Regarding Proposition 4.1D, $\mathrm{var}(\mathbf{y}) \subseteq \mathrm{var}(t)$, but not $\mathrm{var}(\mathbf{y}\eta) \subseteq \mathrm{var}(t\eta)$. Regarding Proposition 4.4A, notice that $\mathbf{x}$ occurs properly in $[\mathbf{0}]$, since $[\mathbf{0}] = [\mathbf{x} - \mathbf{y}][\mathbf{x}/\mathbf{y}]$.

B. Take instead, the quotient system induced by the equations $\mathbf{x} = \mathbf{P_1}(\langle \mathbf{x}, \mathbf{y}\rangle)$, $\mathbf{y} = \mathbf{P_2}(\langle \mathbf{x}, \mathbf{y}\rangle)$, $\mathbf{z} = \langle \mathbf{P_1}(\mathbf{z}), \mathbf{P_2}(\mathbf{z})\rangle$. Let π be the substitution $[\langle \mathbf{x}, \mathbf{y}\rangle/\mathbf{z} + \mathbf{P_1}(\mathbf{z})/\mathbf{x} + \mathbf{P_2}(\mathbf{z})/\mathbf{y}]$. It is easy to see that $\pi^2 = \epsilon$, so that π is invertible, even though it is not variable-valued. Notice that $[\mathbf{z}]$ and $[\mathbf{z}]\pi$ have different *arity*, that is, a different number of free variables. Moreover, the variable $[\mathbf{z}] = [\mathbf{z}]\pi^2$ is an instance of the nonvariable $[\mathbf{z}]\pi$. Let $t = [\mathbf{P_1}(\mathbf{z})]$. Regarding Proposition 4.1B, we have $\text{var}(t\pi) = \text{var}([\mathbf{P_1}(\langle \mathbf{x}, \mathbf{y}\rangle)]) = \{[\mathbf{x}]\}$, but $\cup\{u\pi \mid u \in \text{var}(t)\} = \{[\mathbf{x}], [\mathbf{y}]\}$. This happens even though π is invertible.

C. Take instead, the quotient system induced by the equation $\mathbf{x}*\mathbf{0} = \mathbf{0}$. Regarding Proposition 4.4A, $[\mathbf{x}*\mathbf{y}][\mathbf{z}/\mathbf{x} + \mathbf{0}/\mathbf{y}] = [\mathbf{0}]$, showing that $\mathbf{z}$ occurs properly in $[\mathbf{0}]$, even though $\mathbf{z} \notin \text{var}([\mathbf{0}])$. Regarding Proposition 4.4B, a simple induction shows that if $[\mathbf{0}] = r(s/v)$, with $v \in \text{var}(r)$, then $s = [\mathbf{0}]$. Thus, it is not possible to write $[\mathbf{0}] = r(\mathbf{z}/v)$, with $v \in \text{var}(r)$.

D. Take instead, the quotient system induced by the equation $\mathbf{F}(\mathbf{G}(\mathbf{x})) = \mathbf{F}(\mathbf{G}(\mathbf{0}))$. Regarding Proposition 4.4C, $[\mathbf{x}]$ occurs in $[\mathbf{G}(\mathbf{x})]$ and $[\mathbf{G}(\mathbf{x})]$ occurs in $[\mathbf{F}(\mathbf{G}(\mathbf{0}))]$, but $[\mathbf{x}]$ doesn't occur in $[\mathbf{F}(\mathbf{G}(\mathbf{0}))]$.

4.2. *Unique Quotients.*

A. Starting with first-order term instantiation, take the quotient system induced by the equations $\mathbf{P}(\mathbf{S}(\mathbf{x})) = \mathbf{x}$ and $\mathbf{S}(\mathbf{P}(\mathbf{x})) = \mathbf{x}$. Exercise 6.2 will show that this example has unique quotients (as does [BURC87, Sec. 3, Example E_{12}]). Notice that the variable $[\mathbf{x}]$ is an instance of the nonvariable $[\mathbf{P}(\mathbf{x})]$ (and conversely). Consequently, the conclusion of Proposition 4.2C cannot be strengthened to say that equivalent terms are just renamings of one another. Finally, the term $[\mathbf{P}(\mathbf{x})]$ occurs properly in itself, because $[\mathbf{P}(\mathbf{x})] = [\mathbf{P}(\mathbf{S}(\mathbf{P}(\mathbf{x})))]$. Moreover $[\mathbf{P}(\mathbf{x})]$ is a proper generalization of $[\mathbf{P}(\mathbf{S}(\mathbf{x}))]$, which is a proper generalization of $[\mathbf{P}(\mathbf{x})]$. Thus, this example has both *generalization* cycles and occurrence cycles.

B. Let E be the term congruence generated by the identity $\mathbf{P_1}(\mathbf{x}, \mathbf{y}) = \mathbf{x}$. Then the resulting quotient system has unique quotients.

4.3. *Periodic String Instantiation.* Extend Example 2.3A by allowing infinite, periodic strings as terms, but keep the same set of substitutions. Define instantiation as before. Let **b**, **c** be constant symbols; let u, v be variables. Let s be the infinite term 'u**bcbcbc**...'; let t be 'v**cbcbcb**...'; let δ be the substitution 'v**c**'/u + 'u**b**'/v. Clearly, $s\delta = t$ and $t\delta = s$. But there is no invertible substitution π such that $s\pi = t$. Consequently, the unique-quotient assumption is needed in Proposition 4.2C. (Moreover, it cannot be weakened significantly, because, in this example, substitutions are variable-preserving, by Exercise 5.2D.)

4.4. *Type Strictness.* Counter-examples for Proposition 4.3 are illustrated below. In each case, the intended system is the smallest subsystem of first-order term instantiation to which the diagram applies. (Rigorous methods for constructing subsystems are described in Section 5.)

A. *First-Class Terms.*

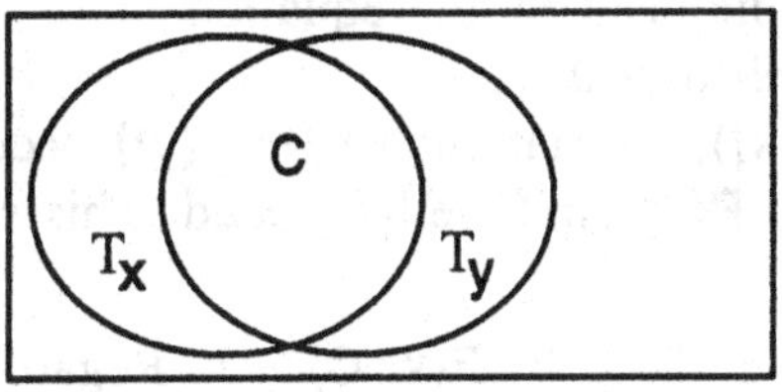

Terms are trivially first-class. But **C** is not weakly restrictable to a common subtype of $T_{\mathbf{x}}$ and $T_{\mathbf{y}}$.

B. *Restrictable Terms.*

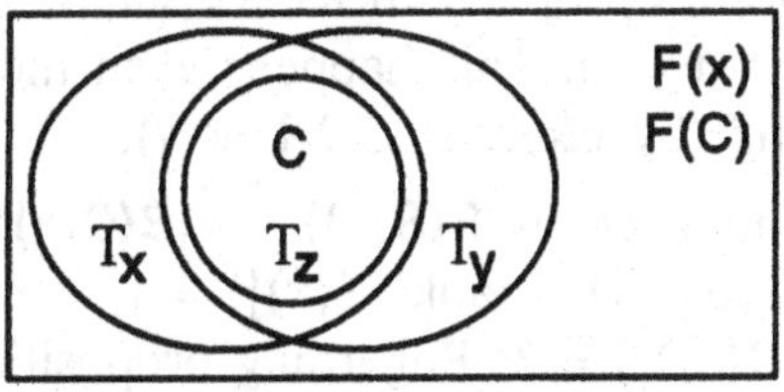

Terms are restrictable. In particular, **x(C/x)** $\in T_{\mathbf{y}}$ implies **x(z/x)** $\in T_{\mathbf{y}}$. But **F(x)**, **F(C)** are not first-class. Variables are not strictly typed.

C. *Weakly Restrictable Terms.*

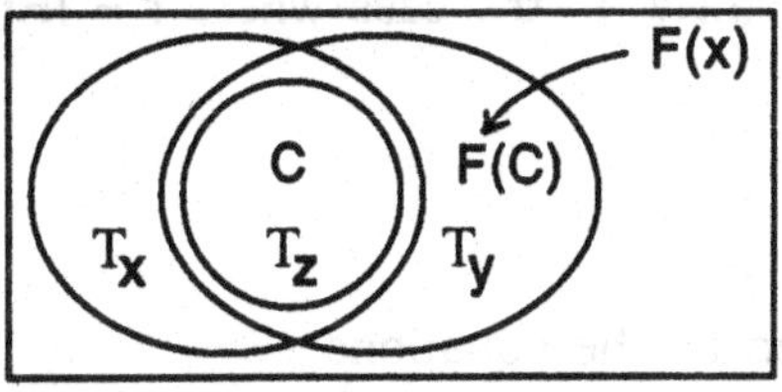

Terms are weakly restrictable. But **F(x)** is not first-class, and **F(x)** is not restrictable to $T_{\mathbf{y}}$. Variables are not strictly typed.

D. *Strict Typing.*

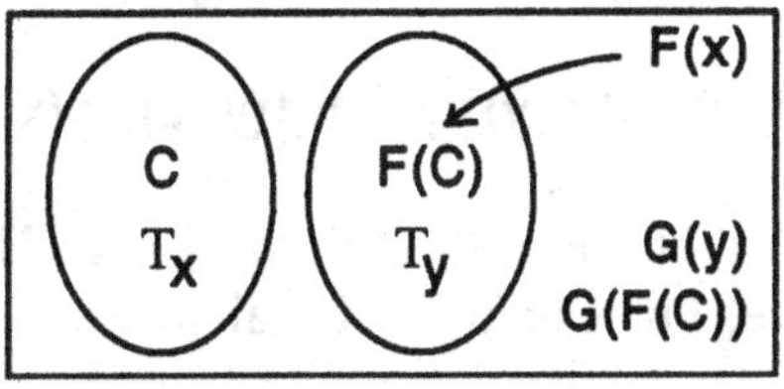

Variables are strictly typed. But **F(x)** is not first-class. **F(x)** is not restrictable to $T_{\mathbf{y}}$. Regarding Proposition 4.4C, **C** occurs in **F(C)**, since **x** $\leq$ **C**, and **F(C)** occurs in **G(F(C))**, since **y** $\leq$ **F(C)**. But **C** does not actually occur in **G(F(C))**, because there is no well-defined term of the form **G(F(**v**))**.

4.5. *Term Occurrences.*

A. Starting with first-order term instantiation, take the quotient system induced by the infinitely many equations $\mathbf{F(Q_1)} = \mathbf{F(Q_2)} = \mathbf{F(Q_3)} = \ldots$. Then the term $[\mathbf{F(Q_1)}]$ is well-founded and has only one interesting generalization, namely [**F(x)**]. But infinitely many terms of the form $[\mathbf{Q_i}]$ properly occur in it.

B. Take instead, the quotient system induced by the infinitely many equations $F_0 = F_1(Q)$; $F_1(x_1) = F_2(x_1, Q)$; $F_2(x_1, x_2) = F_3(x_1, x_2, Q)$; .... Then the term $[F_0]$ is well-founded, since $[Q]$ is the only term that occurs properly in it. However, we do have the following infinite sequence of proper generalizations: $[F_0] > [F_1(x_1)] > [F_2(x_1, x_2)] > \ldots$

C. Take instead, the quotient system induced by the equation $R(x) = G(R(P(x)))$.† Then, for each $n \geq 0$, $[R(P^n(y))] = [G(x)]([R(P^{n+1}(y))]/[x])$, so that each $R(P^{n+1}(y))$ occurs properly in $R(P^n(y))$, showing that $R(x)$ isn't well-founded. This system is still acyclic, however.

D. Take instead, the quotient system induced by $F(x, C) = x$. Regarding Proposition 4.4B, $[x] = [F(x, y)][C/y]$, $[x] \in \mathrm{var}([F(x, y)])$, and $[x][C/y] = [x]$, showing that $[x]$ occurs properly in itself. But we cannot write $[x] = t([x]/u)$, with $u \in \mathrm{var}(t)$, $t \neq u$. This happens even though substitutions are variable-preserving (by Exercise 5.2 below).

E. Take instead, the quotient system induced by $R(S(x)) = A(B(C(x)))$. This system is well-founded, $\mathrm{rank}([S(x)]) = \mathrm{rank}([T(x)]) = 1$, $\mathrm{rank}([R(T(x))]) = 2$, and $\mathrm{rank}([R(S(x))]) = 3$. Regarding proposition 4.5B, we do not have the stronger conclusion that $\mathrm{rank}(r\sigma) \leq \mathrm{rank}(r\tau)$, provided $\mathrm{rank}(v\sigma) \leq \mathrm{rank}(v\tau)$, for all v, in the case where $r = [R(x)]$, $\sigma = [S(x)/x]$, and $\tau = [T(x)/x]$.

4.6. *Fixed Points.*

A. Take the quotient system generated by the equations $f(g(x, z), h(x, y)) = f(y, z)$, $g(f(y, z), h(x, y)) = g(x, z)$, $h(f(y, z), g(x, z)) = h(x, y)$. Let $\sigma = [f(y, z)/x + g(x, z)/y + h(x, y)/z]$. Then

$$\sigma^2 = [f(g(x, z), h(x, y))/x + g(f(y, z), h(x, y))/y + h(f(y, z), g(x, z))/z] = \sigma.$$

But there does not exist an n such that $\mathrm{cdm}(\sigma^n) \cap \mathrm{dom}(\sigma) = \emptyset$. Thus, Condition E of Theorem 4.6 is true, but Condition D fails.

B. *Unique Fixed Points.* Starting with first-order term instantiation, take the quotient system induced by the equations $F(1, 2) = 1$, $F(7, 8) = 7$, $G(1, 2) = 2$, $G(7, 8) = 8$. Then (s/v) has at most one proper fixed point, for all s. But both $[1/x + 2/y]$ and $[7/x + 8/y]$ are improper fixed points of $[F(x, y)/x]$, and are proper fixed points of $[F(x, y)/x + G(x, y)/y]$. Thus, the existence of fixed points for scalar substitutions played an essential role in proving uniqueness in Theorem 4.10C.

† Notice that the equation which generates the congruence relation in this example is essentially just an ordinary recursive definition.

SECTION 5
HOMOMORPHISMS

Quotient homomorphisms provide the necessary tie-in with the literature on equational unification. Full embeddings may be used to improve the behavior of the unification algorithm given in Section 7. Every homomorphism is the composition of a quotient mapping and an embedding. Many of the results presented below are generalizations of propositions from [SCHM88, Sec. 5].

We use superscripts to distinguish components of different instantiation systems, as for example, O^A for the objects of $\mathcal{A}$ or S^B for the substitutions of $\mathfrak{B}$. For any function h and set X, let $h^{\to}(X) = \{h(x) \mid x \in X \cap dom(h)\}$. A function h from O^A to O^B is a *homomorphism* from $\mathcal{A}$ to $\mathfrak{B}$ iff, for each $\sigma \in S^A$, there is a $\tau \in S^B$ such that $dom(\tau) \subseteq h^{\to}(dom(\sigma))$ and $h(t\sigma) = h(t)\tau$, for all t in O^A; notice that τ is uniquely determined; we define $h^{\to}(\sigma)$ to be this unique substitution τ. (The corresponding definition in [SCHM88] is similar, but adds the requirement that $V^B \subseteq h^{\to}(V^A)$.) We say h is *variable-preserving* iff h(v) always a variable.

CONGRUENCES AND QUOTIENT SYSTEMS

An equivalence relation that respects the structure of both terms and substitutions is an *instantiation congruence*. After showing that instantiation congruences induce quotient homomorphisms, we consider further constraints on congruences that guarantee preservation of various classification properties.

Recall that a term congruence is an equivalence relation E such that s E t implies $s\sigma$ E $t\sigma$. A *substitution congruence* is an equivalence relation on terms such that if $v\sigma$ E $v\tau$, for all v, then $t\sigma$ E $t\tau$, for all t.† A term congruence which is also a substitution congruence is an *(instantiation) congruence.* For first-order terms, E is an instantiation congruence iff the corresponding set of equations, $\{s = t \mid s \text{ E } t\}$, is closed under logical inference, that is, is an equational theory. For any binary relation R on terms, there is a smallest instantiation congruence R^* containing R.

An instantiation congruence is *variable-preserving* iff u E v implies u = v. With the possible exception of Example 2.1, the congruence relations discussed in the examples at the end of Section 2 are variable-preserving instantiation congruences. For first-order terms, E is variable-preserving iff the corresponding set of equations is (equationally) consistent.

---

† These two properties are essentially what Gallier and Snyder refer to as *stability* and *monotonicity,* respectively [GALL88].

Let E be an instantiation congruence for an instantiation system $\mathcal{A}$. We may construct a *quotient* structure $\mathcal{A}$/E as follows:

O/E is the set of all E–equivalence classes [t], for t in O.
V/E is $\{[v] \mid$ for all u, u E v implies u = v$\}$.
S/E is the set of all partial functions $[\sigma]$ from V/E to O/E, given by
$[\sigma]([v]) = [v\sigma]$, provided $[v] \neq [v\sigma]$.
$*$/E is given by $[t][\sigma] = [t\sigma]$.

Let $K_E = \{[v] \mid$ for some u, u E v and $u \neq v\}$.† Thus, $K_E = [V]^{\sim} \setminus V/E$, where $[X]^{\sim} = \{[x] \mid x \in X\}$, for any set X. We say that a term t is a *constant* (is *closed*, is a *ground* term [cf GALL88]) iff var(t) = $\emptyset$. Notice that t is a constant iff $t\sigma = t$, for all σ.

Proposition 5.1E below is sharp, since in Example 4.1A, $[\mathrm{var}(\mathbf{x} - \mathbf{x})]^{\sim} = \{[\mathbf{x}]\}$, but $\mathrm{var}([\mathbf{x} - \mathbf{x}]) = \mathrm{var}([\mathbf{0}]) = \emptyset$; this issue is discussed further in Exercise 5.2.

Proposition 5.1. Let E be an instantiation congruence for an instantiation system $\mathcal{A}$.

A. $\mathcal{A}$/E is an instantiation system.
B. $[\sigma|X] = [\sigma]|[X]^{\sim}$.
C. $\mathrm{dom}([\sigma]) = [\mathrm{dom}(\sigma) \setminus \{v \mid v \mathrel{E} v\sigma\}]^{\sim}$.
D. $[\sigma + \tau] = [\sigma] + [\tau]$, provided $\sigma + \tau$ is defined.
E. $[\mathrm{var}(t)]^{\sim} \supseteq \mathrm{var}([t])$.
F. The function [_] from $\mathcal{A}$ onto $\mathcal{A}$/E is a homomorphism.
G. If $v \in K_E$, then $[v] \supseteq \{t \mid v \leq t\}$, so that [v] is a constant.
H. E is variable–preserving iff [_] is; in this case, $[T_u]^{\sim} = T_{[u]}$.

Proof. The proofs of Assertions B through E will be interleaved with the proof of the fact that $\mathcal{A}$/E is an instantiation system. First, to see that $*$/E is well defined, suppose that $[s] = [t]$ and $[\sigma] = [\tau]$. Since $[\sigma] = [\tau]$, $v\sigma \mathrel{E} v\tau$, for all v, so that $t\sigma \mathrel{E} t\tau$, by definition of substitution congruence. But since s E t and E is a term congruence, we have $s\sigma \mathrel{E} t\sigma \mathrel{E} t\tau$. The first three axioms showing that (O/E, S/E, $*$/E) is an action by substitutions on V/E are easily checked.

The restriction axiom follows directly from Assertion B, which follows directly from the definitions involved. Assertion C also follows directly from the definitions involved. One proves Assertion D as follows: Let σ, τ be such that dom([σ]) is disjoint from dom([τ]). Let $X = \mathrm{dom}(\sigma) \setminus \{v \mid v \mathrel{E} v\sigma\}$ and let $Y = \mathrm{dom}(\tau) \setminus \{v \mid v \mathrel{E} v\tau\}$. Then $[\sigma] = [\sigma|X]$, $[\tau] = [\tau|Y]$, $\mathrm{dom}(\sigma|X)$ is disjoint from $\mathrm{dom}(\tau|Y)$, and it is easy to see that $[\sigma] + [\tau] = [\sigma|X + \tau|Y]$. This gives Assertion D.

† It is tempting to consider only variable–preserving congruences where K_E is empty. But notice that the problem of which congruences are variable–preserving contains the undecidable word problem for semigroups as a special case.

Every variable in $\mathcal{A}$/E is equivalent to infinitely many other variables, as a direct consequence of the definition of V/E. Assertion C shows that dom([σ]) is always finite. We next verify that every term t has a finite discriminating set, in fact, $[\text{var}(t)]^{\sim}$ is a discriminating set: Suppose $[v][\sigma] = [v][\tau]$, for each [v] in $[\text{var}(t)]^{\sim}$. Then $v\sigma$ E $v\tau$, for each v in var(t). By definition of substitution congruence, we have $t(\sigma|\text{var}(t))$ E $t(\tau|\text{var}(t))$. But since var(t) discriminates for t, we also have

$$[t][\sigma] = [t\sigma] = [t(\sigma|\text{var}(t))] = [t(\tau|\text{var}(t))] = [t\tau] = [t][\tau].$$

Hence $[\text{var}(t)]^{\sim}$ discriminates for [t]. Assertion E now follows from Theorem 3.7. Regarding Assertion F, it is clear from the construction of $\mathcal{A}$/E that [_] is a homomorphism, where in each case, $[\_]^{\sim}(\sigma) = [\sigma]$. To prove the first part of Assertion G, suppose that u E v, with $u \neq v$, and that $v \leq t$. Then $u = u(t/v)$ E $v(t/v) = t$, so that v E t. Since $*$/E is well defined, we may conclude that for all σ, $[v] = [v\sigma] = [v][\sigma]$, so that [v] is a constant. Finally, the first part of Assertion H follows directly from Assertion G. For the second part, we have,

$$\begin{aligned}[t] \in [T_u]^{\sim} &\text{ iff } u \leq t' \text{ and } t' \text{ E } t, \text{ for some } t\\ &\text{ iff } u\sigma \text{ E } t, \text{ for some } \sigma\\ &\text{ iff } [u] \leq [t] \text{ iff } [t] \in T_{[u]},\end{aligned}$$

where $T_{[u]}$ is well-defined, because [u] is a variable. □

Exercise 5.2. A binary relation R on the terms of $\mathcal{A}$ is *regular* iff var(s) = var(t), whenever s R t [cf BURC87, Sec. 3].

A. If R is regular and substitutions are variable-preserving in $\mathcal{A}$, then the instantiation congruence R^* is regular.
B. If R^* is regular, then R^* is variable-preserving.
C. R^* is regular iff $\text{var}([t]) = [\text{var}(t)]^{\sim}$, for all t.
D. If R^* is regular, then substitutions are variable-preserving in $\mathcal{A}$ iff substitutions are variable-preserving in $\mathcal{A}/R^*$.
E. Consequently, substitutions are variable-preserving in Example 2.2A.

A variable-preserving congruence E on an instantiation system $\mathcal{A}$ is *type-preserving* iff $s \in T_v$, whenever s E t and $t \in T_v$. Thus, E is type-preserving iff $t \in T_v$, whenever $[t] \in T_{[v]}$.

Proposition 5.3. Assume E is variable-preserving and type-preserving.

A. $[T_u]^{\sim} \cap [T_v]^{\sim} = [T_u \cap T_v]^{\sim}$.
B. Terms are weakly restrictable in $\mathcal{A}$ iff they are in $\mathcal{A}$/E.
C. Terms are restrictable in $\mathcal{A}$ iff they are in $\mathcal{A}$/E.

Proof. Assertion A may be proved as follows:

$[t] \in [T_u]^{\sim} \cap [T_v]^{\sim}$ iff
  t E r, $r \in T_u$, t E s, and $s \in T_v$, for some r, s, iff
  $t \in T_u$ and $t \in T_v$ iff $[t] \in [T_u \cap T_v]^{\sim}$.

For Assertion B, first assume terms are weakly restrictable in $\mathcal{A}$; suppose $[t] \in T_{[u]} \cap T_{[v]} = [T_u \cap T_v]^{\sim}$. Since terms are weakly-restrictable in $\mathcal{A}$, we may choose w so that $t \in T_w \subseteq T_u \cap T_v$. But then $[t] \in T_{[w]} \subseteq T_{[u]} \cap T_{[v]}$. The converse is proved similarly.

For Assertion C, first assume terms are restrictable in $\mathcal{A}$; suppose $[t][\sigma] \in T_{[w]}$. Then $t\sigma \in T_w$. Choose η, a weakening for t such that $t\eta \in T_w$ and $v\eta \leq v\sigma$, for each $v \in var(t)$. Then $[\eta]$ is a weakening for $[t]$, $[t][\eta] \in T_{[w]}$, and $[v][\eta] \leq [v][\sigma]$, for each variable $[v] \in var([t])$. To prove the converse, assume terms are restrictable in $\mathcal{A}/E$; suppose $t\sigma \in T_w$. Then $[t][\sigma] \in T_{[w]}$. Let $[\eta]$ be a weakening for $[t]$ such that $[t\eta] \in T_{[w]}$ and $[v\eta] \leq [v\sigma]$, for each $v \in var(t)$. Let $U = \{u \mid [u] \in var([t])\}$. We may assume $dom(\eta) \subseteq U$. Let ρ be a renaming that maps $cdm(\eta) \cap var(t)$ to a set which is disjoint from var(t). Let $\mu = (\eta\rho)|U$. Using the fact that $[\eta]$ is a renaming for $[t]$, one sees easily that μ is a renaming for t. Moreover, $t\mu \in T_w$, because $[t\mu] = [t][\eta][\rho] \in T_{[w]}$. Similarly, one sees that $v\mu$ is equivalent to $v\eta$, which is more general than $v\sigma$, for each v in var(t). □

Exercise 5.4. Let R be a symmetric binary relation on the terms of $\mathcal{A}$.

A. Suppose terms are acyclic and R^* is *simple* in the sense that s does not occur properly in t, whenever s R^* t. Then terms are acyclic in $\mathcal{A}/R^*$. (This result does not extend to R [BURC87, Sec. 3, Example E_3].)

B. Suppose terms are well-founded, and s and t have the same rank whenever s R^* t. Then terms are well-founded in $\mathcal{A}/R^*$. For first-order terms, this result extends to R. (In this case, the above constraint on R is a generalization of *permutativity* [BURC87, LANK77].)

SUBSYSTEMS AND EMBEDDINGS

We first consider *full* subsystems induced by a set of terms. Since not every subsystem is full, we also introduce *type theories* as a mechanism for constructing arbitrary subsystems. Type theories are a generalization of order-sorted signatures. In some cases, the constructed subsystem is isomorphic with the original, and this provides a technique for defining *extensions* as well. One may, for example, extend a system to include constants of every type.

We say that $\mathcal{A}$ is a *subsystem of* $\mathfrak{B}$ iff $O^A \subseteq O^B$, $V^A \subseteq V^B$, $S^A \subseteq S^B$, and $*^A$ is the restriction of $*^B$ to $O^A \times S^A$.

Let A be a set of terms of an instantiation system $\mathfrak{B}$. We say that v is *replete in* A iff $v \in A$ and v is equivalent to infinitely many variables in A. We say A is *instance-closed* iff $t(s/v) \in A$ whenever s, t, $v \in A$ and v is replete in A. If A is instance-closed, we let $\mathfrak{B}|A$ be the following structure:

$O^A = A$,
$V^A = \{v \in A \mid v \text{ is replete in } A\}$,
$S^A = \{\sigma \in S^B \mid \mathrm{dom}(\sigma) \subseteq V^A \text{ and } v\sigma \in O^A, \text{ for all } v \in V^A\}$,
$*^A$ is the restriction of $*^B$ to $O^A \times S^A$.

Proposition 5.5. Let A be an instance-closed set of terms in $\mathfrak{B}$, and let $\mathcal{A} = \mathfrak{B}|A$. Then,

A. $\mathcal{A}$ is an instantiation system.
B. S^A is the largest subset of S^B for which the inclusion of A in O^B is a homomorphism.
C. If $\mathfrak{B}$ is countable, $V^A = V^B \cap A$, and every term in O^B is a renaming of a term in O^A, then $\mathcal{A}$ and $\mathfrak{B}$ are isomorphic.

Proof. To prove Assertion A, we show, first, that $t\sigma \in A$ whenever $t \in A$ and $\sigma \in S^A$: We may assume $\mathrm{dom}(\sigma) \subseteq \mathrm{var}(t)$ and $\sigma = s_1/u_1 + \ldots + s_n/u_n$. By the method of Lemma 3.5, we may let ρ be a renaming substitution in S^A such that $\mathrm{dom}(\rho) = \mathrm{var}(t)$ and $\mathrm{cdm}(\rho)$ is disjoint from $\mathrm{var}(t) \cup \mathrm{cdm}(\sigma)$. If $\rho = v_1/u_1 + \ldots + v_n/u_n$, then $\rho^{\vee}\sigma = s_1/v_1 + \ldots + s_n/v_n$. Moreover, $v_i \notin \mathrm{var}(s_j)$, for $i, j = 1, \ldots, n$. Repeated application of Corollary 3.8D shows that $t\sigma = t\rho\rho^{\vee}\sigma = t(v_1/u_1) \ldots (v_n/u_n)(s_1/v_1) \ldots (s_n/v_n)$, so that $t\sigma \in A$. The now straightforward validation of axioms showing that $\mathcal{A}$ is an instantiation system is omitted.

Assertion B follows directly from the definitions. In particular, if $\iota : O^A \subseteq O^B$ is the inclusion map, then $\iota^{\sim}(\sigma) = \sigma$, for all σ. To prove Assertion C, observe that there is a bijection θ from V^A to V^B such that, in $\mathfrak{B}$, $\theta(v)$ is equivalent to v, for each v. Extend θ to terms in O^A by defining $\theta(t) = t(\theta|\mathrm{var}(t))$. Using Propositions 4.1A and 4.1D, it is straightforward but tedious to verify that θ is an isomorphism; the details are left as an exercise for the reader. □

Subsystems of the form $\mathfrak{B}|A$ are called *full* subsystems. A full subsystem $\mathcal{A}$ of $\mathfrak{B}$ is an *ideal in* $\mathfrak{B}$ iff $t\sigma \in O^A$, whenever $t \in O^A$ and $\sigma \in S^B$. The isomorphism θ constructed in the proof of Proposition 5.5C is an *inner* isomorphism of a system onto a subsystem, meaning that θ satisfies the identity $\theta(t) = t(\theta|\mathrm{var}(t))$, for all terms t. We refer to such a subsystem as an *equivalent* subsystem.

Exercise 5.6.

A. If $\mathcal{A}$ is a subsystem of an untyped instantiation system and, in $\mathcal{A}$, η is a weakening for t, then $\mathrm{var}(t\eta) = \{v\eta \mid v \in \mathrm{var}(t)\}$.
B. Assume $\mathcal{A} \subseteq \mathcal{C} \subseteq \mathfrak{B}$, and $\mathcal{A}$ is a full subsystem of $\mathfrak{B}$. Then $\mathcal{A}$ is a full subsystem of $\mathcal{C}$; moreover, if $v \in V^A$, $t \in O^A$, and, in $\mathcal{C}$, $v \leq t\eta$, and η is weakening for t that is a renaming in $\mathfrak{B}$, then $v \leq t$. We refer to such a weakening as a *slight* weakening in $\mathcal{C}$.
C. If $\mathcal{A}$ is an equivalent subsystem of $\mathfrak{B}$, then $\mathcal{A}$ is a full subsystem of $\mathfrak{B}$, and, moreover, $t \in O^A$ iff $\mathrm{var}(t) \subseteq V^A$, for all $t \in O^B$.

An instantiation system has *finite type intersections* iff whenever $T_u \cap T_v \neq \emptyset$, there exists w such that $T_w = T_u \cap T_v$.

Proposition 5.7. Any instantiation system $\mathcal{A}$ that can be embedded in an untyped system $\mathfrak{B}$ can be fully embedded in a system with finite type intersections.

Proof. Let $\mathcal{T}^A$ be the set of all instance types in $\mathcal{A}$; we write $T^A{}_v$ for the $\mathcal{A}$ instance type of v. Let $\mathcal{T}^\Delta$ be the set of all nonempty finite intersections of elements of $\mathcal{T}^A$. We may assume, by Proposition 5.5C, that $V^B \setminus V^A$ is infinite. Choose a function D from V^B onto $\mathcal{T}^\Delta$ such that if $v \in V^A$, then $D_v = T^A{}_v$, where as, if $v \notin V^A$, then the inverse image of D_v is infinite. We let $V^C = V^B$. If y, $z \in V^C$, let $y \leq^C z$ iff $D_z \subseteq D_y$; this portion of the instance relation for $\mathcal{C}$ determines the variable-valued substitutions in S^C. Let O^C be the set of all terms of the form $r\rho$, where $r \in O^A$, and ρ is a variable-valued substitution in S^C. The instance types of $\mathcal{C}$ are given as follows, $z \in T^C{}_w$ iff $w \leq^C z$; if $t \in O^C \setminus V^C$, then $t \in T^C{}_w$ iff there exist $T_1, \ldots, T_n \in \mathcal{T}^A$ such that $T_1 \cap \ldots \cap T_n \subseteq D_w$, and for each $i = 1, \ldots, n$, there exists $r_i \in T_i$ and variable-valued $\rho_i \in S^C$ such that $t = r_i\rho_i$; we refer to $\{\langle T_i, r_i, \rho_i\rangle \mid i = 1, \ldots, n\}$ as a *witness* set for $t \in T^C{}_w$. This determines S^C. The instantiation operation on $\mathcal{C}$ is the one inherited from $\mathfrak{B}$; this makes sense, because $S^C \subseteq S^B$, because $\mathfrak{B}$ is untyped. We need to show that $\mathcal{C}$ is an instantiation system, that $\mathcal{A}$ is a full subsystem of $\mathcal{C}$, and that $\mathcal{C}$ has finite type intersections.

1. If $\rho, \sigma \in S^C$ and ρ is variable-valued, then $\rho\sigma \in S^C$: we need to show, for each possible w, that $(w\rho)\sigma \in T^C{}_w$. If $w\rho\sigma \in V^C$, this happens because $D_w \supseteq D_{w\rho} \supseteq D_{w\rho\sigma}$. If $w\rho\sigma \notin V^C$, then any witness set for $w\rho\sigma \in T^C{}_{w\rho}$ is also a witness set for $w\rho\sigma \in T^C{}_w$.

2. If $t \in O^C$ and $\sigma \in S^C$, then $t\sigma \in O^C$: let $t = r\rho$, with $r \in O^A$ and ρ variable-valued in S^C. We know $\rho\sigma \in S^C$. From the definitions of O^C and S^C, we may construct $\tau \in S^A$ and $\eta \in S^C$, with η variable-valued, so that $v\rho\sigma = v\tau\eta$, for each $v \in var(r)$. Consequently, $t\sigma = (r\tau)\eta \in O^C$.

3. If $s, t \in O^A$, $\sigma \in S^C$, and $s\sigma = t$, then, in $\mathcal{A}$, $s \leq t$ (so that $\mathcal{A}$ will be a full subsystem of $\mathcal{C}$): As in the previous step, we have $s\tau\eta = t$, with $\tau \in S^A$, $\eta \in S^C$ and η variable-valued. Moreover, we may assume $dom(\tau) \subseteq var(s)$, $dom(\eta) \subseteq var(s\tau)$, and $cdm(\eta) \subseteq V^C \setminus V^A$. We may also take η to be a weakening for $s\tau$: if $v\eta = w$, for all v in some $K \subseteq var(s\tau)$, with K having more than one element, pick $v \in K$, and pick $r \in D_{v\eta}$. Then $r \in D_u \subseteq O^A$, for each $u \in K$. Assume, by renaming variables, if necessary, that $var(r) \cap dom(\eta) = \emptyset$. Let κ be the substitution in S^A that maps each $v \in K$ to r. We know $v\eta \notin var(t)$, so that $t(r/v\eta) = s\tau\eta(r/v\eta) = s(\tau\kappa(\eta|(V^C \setminus K)))$. Replacing τ with $\tau\kappa$, η with $\eta|(V^C \setminus K)$, and following up with a trivial induction on the size of η, we obtain a factorization where η is is a

weakening for $s\tau$. Finally, in $\mathfrak{B}$, η is a renaming for $s\tau$, since $\mathfrak{B}$ is untyped, and $\mathrm{var}(t) = \{w\eta \mid w \in \mathrm{var}(s\tau)\}$, by Proposition 4.1A. Hence, $\mathrm{dom}(\eta) \cap \mathrm{var}(s\tau) = \emptyset$, so that $s\tau = t$, and, in $\mathcal{A}$, $s \leq t$.

4. S^C is closed under composition, and thus $\mathcal{C}$ is an instantiation system. We need to show that if $\sigma, \tau \in S^C$, then $v \leq v\sigma\tau$, for all v. If $v\sigma$ and $v\sigma\tau$ are variables, this happens because $D_{v\sigma\tau} \subseteq D_{v\sigma}$. If $v\sigma$ is a variable and $v\sigma\tau$ is not, this happens because any witness set for $v\sigma\tau \in T^C{}_{v\sigma}$ is a witness set for $v\sigma\tau \in T^C{}_v$. If $v\sigma$ is not a variable, let $\{\langle T_i, r_i, \rho_i\rangle \mid i = 1, \ldots, n\}$ be a witness set for $v\sigma \in T^C{}_v$, so that $v\sigma\tau = r_i\rho_i\tau$, for each i. Using the fact that each $u\rho_i\tau$ belongs to O^C, we may refactor $\rho_i\tau$, so that $\rho_i\tau \equiv \gamma_i\eta_i\ [\mathrm{var}(r_i)]$, with each γ_i in S^A and each η_i variable-valued with $\mathrm{dom}(\eta_i) \subseteq \mathrm{var}(r_i\gamma_i)$. If $v\sigma\tau$ is not a variable, we have, $v\sigma\tau = r_i\rho_i\tau = r_i\gamma_i\eta_i$, with each $r_i\gamma_i \in T_i$, so that $\{\langle T_i, r_i\gamma_i, \eta_i\rangle \mid i = 1, \ldots, n\}$ is a witness set for $v\sigma\tau \in T^C{}_v$. If $v\sigma\tau$ is a variable, say $v\sigma\tau = w$, we shall show $D_w \subseteq D_v$, by showing $D_w \subseteq T_i$, for each i. To do this, first refactor $\gamma_i\eta_i$ so that $\mathrm{cdm}(\eta_i) = \{w\}$: repeatedly select a nonempty set $K \subseteq \mathrm{dom}(\eta_i)$ with $\mathrm{cdm}(\eta_i|K) = \{y\}$ with $y \neq w$, and, as in Step 3, modify γ_i and restrict η_i to $V \setminus K$. Finally, pick any $t \in D_w$. Since $\mathrm{dom}(\eta) \subseteq V^A$, the functional composition $\eta_i \circ (t/w)$ belongs to S^A. We have, $r_i\gamma_i \leq r_i\gamma_i(\eta_i \circ (t/w)) = v\sigma\tau(t/w) = t$, showing that t belongs to T_i.

5. The system $\mathcal{C}$ has finite type intersections: Pick $u, v \in V^C$, with $T^C{}_u \cap T^C{}_v \neq \emptyset$. Then, $T^A{}_u \cap T^A{}_v \neq \emptyset$, so that $D_u \cap D_v \neq \emptyset$, and we may choose $w \in V^C$ so that $D_w = D_u \cap D_v$; it suffices to show that $T^C{}_w = T^C{}_u \cap T^C{}_v$. For any variable x, $x \in T^C{}_w$ iff $D_x \subseteq D_w = D_u \cap D_v$ iff $x \in T^C{}_u \cap T^C{}_v$. For any nonvariable t, if Ω is a witness set for $t \in T^C{}_w$, then Ω is a witness set for both $T^C{}_u$ and $T^C{}_v$. Conversely, if $t \in T^C{}_u \cap T^C{}_v$, let Σ and Γ be witness sets for $t \in T^C{}_u$ and $t \in T^C{}_v$, respectively. Then $\Sigma \cup \Gamma$ is a witness set for $t \in T^C{}_w$. □

A *type theory* for an instantiation system $\mathfrak{B}$ is given by a quadruple of the form $(U, \mathcal{T}, \sqsubseteq, \mathsf{E})$, where U is a subset of V^B, $\mathcal{T}$ is an arbitrary set of *sorts*, $\sqsubseteq$ is a partial order on $\mathcal{T}$, and E is a subset of $O^B \times \mathcal{T}$. We let $T, T_1, T_2, \ldots$ vary over elements of $\mathcal{T}$, write $t \mathrel{\mathsf{E}} T$, to mean $\langle t, T\rangle \in \mathsf{E}$, refer to the elements of E as *typing axioms,* and refer to the terms occurring in typing axioms as *primitive* terms. We require that type theories satisfy the following axioms:

ω-repleteness: for each $u \in U$, there are infinitely many $v \in U$ such that $u \mathrel{\mathsf{E}} T$ iff $v \mathrel{\mathsf{E}} T$, for all T.

v-regularity: for each $u \in U$ there is a unique smallest $T \in \mathcal{T}$ for which $u \mathrel{\mathsf{E}} T$; we refer to T as the *declared type* of u.

type inheritance: whenever $t \mathrel{\mathsf{E}} T_1$, $T_1 \sqsubseteq T_2$, and T_2 is the declared type of v, we have $t \in T_v$.

The ω–repleteness axiom is necessary to guarantee that the induced subsystem is an instantiation system. The v–regularity axiom guarantees that type–inclusion is consistent with the inclusion relation on instance types. Type inheritance rules out false substitutions that are allowed by the theory, even though they don't belong to $\mathfrak{B}$.

In specifying a type theory, we usually violate ω–repleteness, intending that the theory be extended by adding infinitely many variables of each declared type to U. We may also omit U from the theory, intending that U is the set of all primitive terms belonging to V^B.

We simultaneously define a minimal set of *type-preserving* substitutions and a minimal *of-type* relation between terms and types as follows: If $t \in T$, then t is of type T; if t is of type T_1 and $T_1 \sqsubseteq T_2$, then t is of type T_2; if σ is type–preserving, then tσ is of type T whenever t is; if $\mathrm{dom}(\sigma) \subseteq U$ and if vσ is of type T whenever v is, then σ is type–preserving. A term is *well–typed* iff it is of type T, for some T.

Any order–sorted signature $(\mathcal{T}, \sqsubseteq, \mathcal{F})$ has an *associated* type theory $(U, \mathcal{T}, \sqsubseteq, \in)$ given as follows. Each $T \in \mathcal{T}$ is associated with infinitely many variables whose declared type is T. Each function symbol f and signature $f : T_1 \times \dots \times T_n \to T$ is associated with a typing axiom of the form $f(x_1, \dots, x_n) \in T$, where each x_i is of declared type T_i. Constants and variables are also associated with typing axioms, in the obvious way.

Exercise 5.8. Let $(U, \mathcal{T}, \sqsubseteq, \in)$ be an ω–replete (extension of a) type theory for an instantiation system $\mathfrak{B}$.

A. The well–typed terms, variables in U, type–preserving substitutions, and inherited instantiation operation form a subsystem of $\mathfrak{B}$, which we refer to as the *induced* subsystem.
B. If T is the declared type of a variable, then the set of all terms of type T is an instance type of the induced subsystem.
C. An order–sorted signature and its associated type theory induce the same subsystems of first–order term instantiation.
D. If $w \in V^B \setminus U$ and there is a typing axiom of the form $w \in T$, for some $T \in \mathcal{T}$, then w is a constant in the induced subsystem.
E. If $w \in U$, but there is no typing axiom of the form $t \in T$, where t is a nonvariable and T is the declared type of w, then w is a *parameter* of the induced subsystem, in the sense that every instance of w is a variable.

Let $\mathcal{A}$ be any instantiation system. For each $v \in V^A$, let $T^A{}_v$ be the set of all $\mathcal{A}$–instances of v. Let $\mathcal{T} = \{T^A{}_v \mid v \in V^A\} \cup \{O^A\}$. Let $\sqsubseteq$ be the subset relation on $\mathcal{T}$. Let $\in$ be the set of all $\langle t, T\rangle$ such that $t \in T \in \mathcal{T}$.
Then $(V^A, \mathcal{T}, \sqsubseteq, \in)$ is the *intrinsic* type theory for $\mathcal{A}$. If $\mathcal{A}$ is a subsystem of $\mathfrak{B}$,

and if $(U, \mathcal{T}, \sqsubseteq, \mathrm{E})$ is a type theory that contains the intrinsic theory for $\mathcal{A}$, then the subsystem of $\mathcal{B}$ induced by $(U, \mathcal{T}, \sqsubseteq, \mathrm{E})$ is referred to as the *extension of $\mathcal{A}$ induced by* $(U, \mathcal{T}, \sqsubseteq, \mathrm{E})$.

Exercise 5.9. Assume $\mathcal{A}$ is a subsystem of $\mathcal{B}$ and $\mathcal{C}$ is the extension of $\mathcal{A}$ induced by a type theory containing the intrinsic theory for $\mathcal{A}$.

A. $\mathcal{A}$ really is a subsystem of $\mathcal{C}$, and if $\mathcal{C}$ is induced by the intrinsic theory itself, then $\mathcal{A} = \mathcal{C}$.

B. Assume $\mathcal{A}$ is an equivalent subsystem of $\mathcal{B}$, and every primitive nonvariable of $\mathcal{C}$ belongs to O^A. Then every term in $\mathcal{C}$ is of the form $t\eta$, where $t \in O^A$, and η is a slight weakening for t; t is unique up to a renaming of variables. Moreover, every substitution in $\mathcal{C}$ is of the form $\tau\rho|X$, where τ is in $\mathcal{A}$ and ρ is a slight weakening. Moreover, if $v \in V^C$, η is a slight weakening for s, and $v \leq s\eta \in O^C \setminus V^C$, then $v \leq s$.

Proposition 5.10. Any instantiation system can be fully embedded in one that has infinitely many constants of every type [SCHM88, Lemma 5.11].

Proof. Let $\mathcal{A}$ be an arbitrary instantiation system. Proposition 5.5C shows how to find an isomorphism from $\mathcal{A}$ to a full subsystem $\mathcal{A}'$ that fails to contain infinitely many variables of each type. One may then extend the inverse isomorphism to one that maps all of $\mathcal{A}$ to a full extension $\mathcal{B}$ such that every variable in V^A is equivalent to infinitely many variables in $V^B \setminus V^A$. Now consider the extension of the intrinsic type theory for $\mathcal{A}$ obtained by adding all typing axioms of the form $u \in T^A{}_v$, where $u \in V^B \setminus V^A$, $v \in V^A$, and u is equivalent to v (but not adding u to U). Then the induced extension of $\mathcal{A}$ contains infinitely many constants of every type, by Exercise 5.8D. □

ARBITRARY HOMOMORPHISMS

The remaining task for this section is to put together what has already been learned about subsystems, quotient systems, and homomorphisms. The main point is that there are nontrivial examples of bijective homomorphisms that are not isomorphisms. These exist because homomorphisms are officially defined only on terms, and substitutions are regarded as part of the structure of an instantiation system. In this regard, the category of instantiation systems differs significantly from categories of many-sorted algebras.

Theorem 5.11. Let h be a homomorphism from $\mathcal{A}$ to $\mathcal{B}$.

A. If $h(u) = h(v)$ and $u \neq v$, then $h(v) = h(v\sigma)$, for all σ.

B. If $h|V^A$ is one-to-one, then h is variable-preserving.

C. The relation $=_h$ given by $s =_h t$ iff $h(s) = h(t)$ is an instantiation congruence.

D. h may be uniquely expressed as the composition of the three homomorphisms, $h = [\_] \circ e \circ \iota : \mathcal{A} \twoheadrightarrow \mathcal{A}/{=_h} \rightarrowtail\!\!\!\twoheadrightarrow \mathfrak{B}|H \subseteq \mathfrak{B}$, where $[\_]$ is the quotient map induced by $=_h$, e is bijective, H is the range of h, and $\iota : H \subseteq O^B$ is the inclusion map.

Proof. Regarding Assertion A, suppose $h(u) = h(v)$ and $u \neq v$. We have, $h(v) = h(u) = h(u(v\sigma/v)) = h(u)\breve{h}(v\sigma/v) = h(v)\breve{h}(v\sigma/v) = h(v(v\sigma/v)) = h(v\sigma)$.

For Assertion B, Pick any v; pick u equivalent to v with $u \neq v$. Then $h(v) \neq h(u) = h(v(u/v)) = h(v)\breve{h}(u/v)$. Since $\mathrm{dom}(h(u/v)) \subseteq \{h(v)\}$ and $h(u/v)$ is not the identity, we conclude that $h(v) \in \mathrm{dom}(h(u/v))$ and $h(v)$ is a variable.

Regarding Assertion C, the equivalence relation $=_h$ is a term congruence directly because h is a homomorphism. To see that $=_h$ is a substitution congruence, suppose $h(v\sigma) = h(v\tau)$, for all v. Then $\breve{h}(\sigma) = \breve{h}(\tau)$. Consequently, for all t, $h(t\sigma) = h(t)\breve{h}(\sigma) = h(t)\breve{h}(\tau) = h(t\tau)$, as required.

Finally, regarding Assertion D, $[\_]$ is a homomorphism by Proposition 5.1G; e is uniquely determined by the equation $e([t]) = h(t)$. Proposition 5.5B shows that ι is a homomorphism, since H is instance closed, this being established via the calculation $h(t)(h(s)/h(v)) = h(t)\breve{h}(s/v) = h(t(s/v))$. Finally, simple, omitted calculations show that $e([\sigma]) = h(\sigma)$, and, thus, e is a homomorphism. □

Relative to the factorization given in Theorem 5.11D, we say h is an *embedding* iff $[\_]$ is the identity, and is *full* iff e is an isomorphism. Thus, h is full iff $h(s) \leq h(t)$ implies $s =_h s' \leq t' =_h t$, for some s', t'. If $\mathcal{A}$ is a subsystem of $\mathfrak{B}$, then $\mathcal{A}$ is a full subsystem iff the identity $\iota : \mathcal{A} \to \mathfrak{B}$ is full. Finally, we define h to be a *quotient* homomorphism iff h is both full and onto. Thus, h is a quotient homomorphism iff h factors through the quotient system $\mathcal{A}/{=_h}$. In this case, we say h is *type-preserving* or *regular* according to whether $=_h$ is.

Exercise 5.12. Suppose $\mathfrak{B}$ is a full subsystem of $\mathcal{C}$ and E is an instantiation congruence on $\mathfrak{B}$. Let E' be the smallest congruence on $\mathcal{C}$ that contains E. Let e be the homomorphism from $\mathfrak{B}/E$ to $\mathcal{C}/E'$ given by $e([s]_E) = [s]_{E'}$, for all $s \in O^A$.

A. If $\mathfrak{B}$ is an ideal in $\mathcal{C}$, then $\mathfrak{B}/E$ is an ideal in $\mathcal{C}/E'$.

B. If E is variable-preserving and $\mathfrak{B}$ is an equivalent subsystem of $\mathcal{C}$, then E' is variable-preserving and $\mathfrak{B}/E$ is a full subsystem of $\mathcal{C}/E'$. (See Example 5.1C regarding sharpness.)

$$\begin{array}{ccc} \mathcal{C} & \twoheadrightarrow & \mathcal{C}/E' \\ \cup\!| & & \uparrow e \\ \mathfrak{B} & \twoheadrightarrow & \mathfrak{B}/E \end{array}$$

COUNTEREXAMPLES

5.1. *Quotients and Embeddings.*

A. Let $\mathcal{A}$ be the order-sorted instantiation system with just two unrelated types T_1 and T_2, with type theory

$$\mathbf{x}_1, \mathbf{y}_1 \in T_1; \quad \mathbf{x}_2, \mathbf{y}_2 \in T_2;$$
$$\mathbf{H}(\mathbf{x}_1, \mathbf{y}_1) \in T_1; \quad \mathbf{H}(\mathbf{x}_2, \mathbf{y}_2) \in T_2.$$

Let E be the congruence generated by the equations $\mathbf{H}(\mathbf{x}_1, \mathbf{y}_1) = \mathbf{x}_1$; $\mathbf{H}(\mathbf{x}_2, \mathbf{y}_2) = \mathbf{y}_2$. Then E is variable-preserving.

B. Let $\mathfrak{B}$ be an extension of $\mathcal{A}$ that contains variables $\mathbf{x}_3$, $\mathbf{y}_3$, ... that are simultaneously of type T_1 and of type T_2. Let E′ be any congruence on $\mathfrak{B}$ that contains E. Then $\mathbf{x}_3$ E′ $\mathbf{H}(\mathbf{x}_3, \mathbf{y}_3)$ E′ $\mathbf{y}_3$, so that E′ cannot be variable-preserving.

C. Let $\mathcal{C}$ be first-order term instantiation. Let $\mathfrak{B}$ be the subsystem of $\mathcal{C}$ given by $\mathbf{T_z} \subseteq \mathbf{T_x}$; $\mathbf{z} \in \mathbf{T_z}$; $\mathbf{x} \in \mathbf{T_x}$; $\mathbf{w}$, $\mathbf{c}$, $\mathbf{f(x)} \in \mathbf{T}$. Let E be the congruence on $\mathcal{C}$ generated by $\mathbf{f(z)} = \mathbf{c}$. Consider the quotient system $\mathfrak{B}$/E. Regarding Proposition 4.1A, if $t = [\mathbf{f(x)}]$ and $\eta = [\mathbf{z}/\mathbf{x}]$, then $\emptyset = \mathrm{var}(t\eta) \neq \{v\eta \mid v \in \mathrm{var}(t)\} = \{[\mathbf{z}]\}$, even though η is a weakening for t. Regarding Exercise 5.12, if E′ is the smallest congruence for $\mathcal{C}$ that contains E, then the mapping e: $\mathfrak{B}$/E -> $\mathcal{C}$/E′ is not an embedding because $\mathfrak{B}$/E is not a subsystem of any untyped system, by Exercise 5.6A, since $\mathcal{C}$/E′ is untyped. This example isn't as sharp as it could be, however, because $\mathfrak{B}$ is not a full subsystem of $\mathcal{C}$.

5.2. *Nonrestrictable Terms.*

A. Starting with first-order term instantiation, take the subsystem induced by the type theory $(\mathcal{T}, \subseteq, \mathcal{A})$, where $\mathcal{T} = \{\mathrm{R}, \mathrm{C}\}$, $\mathrm{R} \subseteq \mathrm{C}$, and $\mathcal{A}$ is:

$$\mathbf{x}, \mathbf{x}_1, \mathbf{x}_2 \in \mathrm{R}; \qquad \mathbf{z}, \mathbf{z}_1, \mathbf{z}_2 \ldots \in \mathrm{C};$$
$$\mathbf{Im(z)} \in \mathrm{R}; \qquad \mathbf{i}, \mathbf{z}_1 - \mathbf{z}_2, \mathbf{z}_1 * \mathbf{z}_2 \in \mathrm{C};$$
$$\mathbf{z} - \mathbf{i} * \mathbf{Im(z)} \in \mathrm{R}.$$

Let t be the term $\mathbf{z}_1 - \mathbf{z}_2$; let $\sigma = \mathbf{z}/\mathbf{z}_1 + \mathbf{i}*\mathbf{Im(z)}/\mathbf{z}_2 + (\mathbf{z} - \mathbf{i}*\mathbf{Im(z)})/\mathbf{x}$. Regarding Proposition 4.3, terms are first class and weakly restrictable, but not restrictable: $t\sigma = \mathbf{z} - \mathbf{i}*\mathbf{Im(z)} \in \mathrm{R}$, but $t\eta \notin \mathrm{R}$, whenever $t\eta \leq t\sigma$ and η is a weakening for t. Hence, this is not an order-sorted system in the sense of Exercise 3.12A. Notice that σ is a unifier for $\mathbf{x}$ and $\mathbf{z}_1 - \mathbf{z}_2$. It is neither more nor less general than the unifier $\tau = (\mathbf{x}_1 - \mathbf{x}_2)/\mathbf{x} + \mathbf{x}_1/\mathbf{z}_1 + \mathbf{x}_2/\mathbf{z}_2$.

B. Take instead, the order-sorted subsystem induced by the typing axioms

$\mathbf{x}, \mathbf{x}_1, \mathbf{x}_2 \in \mathbf{R}$; $\mathbf{z}, \mathbf{z}_1, \mathbf{z}_2 \in \mathbf{C}$;

$\mathbf{Re(z)}, \mathbf{Im(z)} \in \mathbf{R}$; $\mathbf{i}, \mathbf{z}_1 - \mathbf{z}_2, \mathbf{z}_1 * \mathbf{z}_2 \in \mathbf{C}$.

Then take the quotient system induced by the equation $\mathbf{Re(z)} = \mathbf{z} - \mathbf{i} * \mathbf{Im(z)}$. Again, terms are first class and weakly restrictable, but not restrictable: $[\mathbf{z} - \mathbf{z}_2][\mathbf{i} * \mathbf{Im(z)}/\mathbf{z}_2] = [\mathbf{Re(z)}] \in \mathbf{R}$, but $[\mathbf{z} - \mathbf{z}_2][v/\mathbf{z}_2] \notin \mathbf{R}$, for any v different from z.

5.3. *Fixed Points and Unifiers.*

A. Starting with first-order term instantiation, take the quotient system induced by $\mathbf{F(Q(x, y))} = \mathbf{Q(x, y)}$. This system isn't well-founded because $[\mathbf{Q(x, y)}]$ occurs properly in itself. Let $\sigma = [\mathbf{F(z)}/\mathbf{z}]$, and let $\tau = [\mathbf{Q(x, y)}/\mathbf{z}]$. Regarding Theorem 4.6, τ is a proper fixed point of σ that is not of the form σ^n. Notice that τ is a most general unifier for $\mathbf{z}$ and $\mathbf{z}\sigma$. Let $\zeta = [\mathbf{Q(y, y)}/\mathbf{z}]$. Then ζ is a unifier of $\mathbf{y}$ and $\mathbf{y}\sigma$ which is not most general and is not of the form $\tau\eta$, for any η.

B. Proceed as above, but use, instead, the subsystem generated by the declarations $T_1, T_2 \in \mathcal{T}$;

$\mathbf{x}, \mathbf{y} \in T_1$; $\mathbf{z}, \mathbf{Q(x, y)}, \mathbf{F(z)} \in T_2$.

As in Example 3A, τ is a fixed point for σ. Regarding Exercise 4.7, there is no fixed point τ' of σ such that $\mathrm{cdm}(\tau') \subseteq \mathrm{cdm}(\sigma)$, since there are no constants of type T_1.

SECTION 6
CONSTRUCT BASES

A plausible starting point in finding implementations for terms is to identify a set of *basic syntactic constructs* that forms a *basis* from which all terms may be generated. We shall describe construct bases axiomatically, since, in general, there is no best way to derive a basis from the underlying structure of the instantiation system.

We give some examples and nonexamples of construct bases, investigate an interesting tie-in between constructs and unique quotients, and identify some elementary properties of constructs that rely only on the fact that every term is an instance of a construct. *Tree-implementations* are then introduced as homomorphisms from finite-tree systems to arbitrary instantiation systems. As an application, Example 6.3 gives a tree-implementation for first-order term- and formula-schemas modulo α-instantiation. Finally, constructs are used in defining term *weights*; properties of term weights play a key role in the study of completeness and computational complexity.

An immediate consequence of our approach is that there are no multiple-arity constructs. Below, Examples 6.5 and 6.6 investigate possible relationships between instantiation and multiple-arity operators.

A term is *finitely generated over* a set C iff it belongs to the smallest set X such that $C \subseteq X$, $V \subseteq X$, and $r(s/v) \in X$ whenever $r, s \in X$ and $v \leq s$. A *construct basis* for an instantiation system is a set C of nonvariable terms, called *constructs,* with the property that every term is finitely generated over C.† The symbols c, d, c_1, d_1, ... will always vary over constructs. Trivially, if $C = O \setminus V$, then C is a construct basis.

If C is a construct basis for $\mathcal{A}$ and E is a variable-preserving congruence relation on $\mathcal{A}$, then the set of all E–equivalence classes of constructs in C is a construct basis for $\mathcal{A}$/E. The set of all primitive terms of a type theory is a construct basis for the induced subsystem. As a special case, the type theory associated with an order-sorted instantiation system provides a construct basis for that order-sorted instantiation system. Notice that if $\mathcal{A}$ is a subsystem of $\mathfrak{B}$ and C is a construct basis for $\mathfrak{B}$, then $A \cap C$ need not be a construct basis for $\mathcal{A}$, since not every term in A is an instance of a construct that belongs to A.

For any instantiation system $\mathcal{A}$, the *minimal* nonvariables of $\mathcal{A}$ are those nonvariable terms r such that if $t \leq r$, then t is a variable or $r \leq t$. In the case

† In Nerode's terminology, C is a "generating set" for the system [NERO59].

of first-order term instantiation, the minimal nonvariables consist of all individual constants, together with all terms of the form $f(v_1, ..., v_n)$. They form a construct basis, which we refer to as the *minimal-nonvariable* basis.

We say that terms are *generalization well-founded* iff there are no infinite descending sequences of nonvariable terms (that is, where each t_i is an instance of t_{i+1} but not conversely). In this case, every nonvariable is an instance of a minimal nonvariable. Example 4.5B violates this condition, and the term $[F_0]$ is not an instance of a minimal nonvariable. For order-sorted instantiation systems, there are no infinite descending sequences of nonvariable terms, provided the associated order-sorted signature contains no infinite ascending sequences of types. Meseguer et al. refer to such signatures as *Noetherian* [MESE87].

In Examples 2.2A and 4.3 (string instantiation), the minimal nonvariables consist of constants of the form 'a' and all two-element strings of the form 'uv.' In both examples, every term is an instance of a minimal nonvariable. In Example 2.2, the minimal nonvariables form a construct basis. But, in Example 4.3, they do not, because some terms (namely, the infinite periodic strings) are not finitely generated. If terms are both well-founded and generalization well-founded, then the minimal nonvariables form a basis.

Even if the set of minimal nonvariables is a basis, it may be an undesirable basis: In Example 2.3C, the minimal nonvariables consist of the usual constructs for first-order terms, together with those *term-free* formulas whose terms (and subterms) are either free variables or contain bound variables, as for example, $(\forall x)(x < g(y, f(x)))$. See Example 6.3 below for further discussion.

Exercise 6.1. Consider the minimal-nonvariable basis for a system in which terms are well-founded, generalization well-founded, and have unique quotients. Suppose E is a congruence such that whenever $c\sigma$ E $c\tau$, we have $v\sigma$ E $v\tau$, for all v in var(c). (Szabó refers to such congruences as *Ω-free* [BURC87, SZAB82].)

A. If $s\sigma$ E $s\tau$, then $v\sigma$ E $v\tau$, for all $v \in$ var(s), by term induction.
B. The resulting quotient system has unique quotients.
C. In particular, Example 4.2A has unique quotients, since, for example, if $[S(s)] = [S(t)]$, then $[s] = [P(S(s))] = [P(S(t))] = [t]$.
D. Treat first-order terms and quantifier-free formulas as an order-sorted system; then the quotient system induced by $(x < y)$ iff $(y > x)$ has unique quotients (cf [BURC87; Lemma 3.5(i)]).
E. Example 2.3C also has unique quotients.

BASIC PROPERTIES OF CONSTRUCTS

Some results in this section and the next depend on properties of terms and constructs that cannot be assumed in general. *Unique* constructs and *skeletal* construct bases are needed for the discussion of tree-implementations. *Potential* constructs, *least-general* constructs, and *rigid* terms are relevant to the efficiency of unification algorithms.

For a given construct basis, we say that c is a,

construct *for* t iff $c \leq t$.

potential construct *for* t iff c is a construct for some instance of t.

least-general construct for t iff whenever d is a construct for t, $d \leq c$.

unique construct for t iff t is a nonvariable and whenever $c \leq t$ and $d \leq t$, d is a renaming of c.

A nonvariable t *has unique* constructs iff every construct for t is unique. If every nonvariable has unique constructs, then every construct is a least-general construct. For a given instantiation system and construct basis, we say that *terms have unique constructs* iff no variable is an instance of a nonvariable and every nonvariable has unique constructs. Finally, a *skeletal* construct basis is one in which no construct is a renaming of any other construct. Obviously, every construct basis contains a skeletal basis.

It can easily happen that c is a potential construct for t without being a construct for t, even for subsystems of first-order term instantiation: In Example 5.2A, take as constructs, the terms that occur in the typing axioms. Then the term $\mathbf{z} - \mathbf{Im}(\mathbf{z})$ is a construct for an instance of $\mathbf{z}_1 - \mathbf{z}_2$, but not for $\mathbf{z}_1 - \mathbf{z}_2$ itself.

In Example 4.2A with minimal-nonvariable constructs, terms do not have unique constructs, because the terms $[\mathsf{P}(\mathsf{v})]$ and $[\mathsf{S}(\mathsf{v})]$ are equivalent minimal nonvariables that are not simply renamings of one another. In Example 2.2 with minimal-nonvariable constructs, terms have unique constructs even though they fail to have unique quotients.

A term t is *rigid* (for a given construct basis) iff whenever $c\alpha = t\sigma$, there exists γ such that $c\gamma = t$ and $(\gamma\sigma)|var(c) = \alpha|var(c)$; we abbreviate the latter condition as $\gamma\sigma \equiv \alpha\ [var(c)]$.

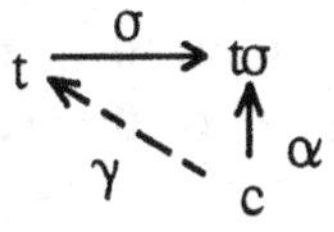

Proposition 6.2. For any construct basis, we have,

A. If t is rigid, then every potential construct for t is a construct for t.
B. If t is rigid, so is $t\sigma$.
C. If c is rigid, then no construct is a proper instance of c.
D. If c is a rigid construct for t, then c is a least-general construct for t.
E. If every nonvariable has unique constructs and unique quotients, then every nonvariable is rigid.

Proof. Assertion A is immediate from the definition. For assertion B, suppose $c\alpha = (t\sigma)\tau$. Choose β so that $c\beta = t$ and $\beta(\sigma\tau) \equiv \alpha$ [var(c)]. Taking $\gamma = \beta\sigma$, we have $c\gamma = t\sigma$ and $\gamma\tau \equiv \alpha$ [var(c)]. For Assertion C, suppose $c\sigma = d$. Since $d\epsilon = c\sigma$, we may choose γ so that $d\gamma = c$ (and $\gamma\sigma \equiv \epsilon$ [var(c)]). Assertion D follows immediately from Assertion C. Finally, for Assertion E, suppose $c\alpha = t\sigma$, with t a nonvariable. Let d be a construct for t. Since d is a construct for $t\sigma$, it is a renaming of c. Hence, c is a construct for t itself. Let $\gamma = t/c$. Then, $c\gamma = t$ and $\alpha = t\sigma/c = ((t/c)\sigma)|var(c) = (\gamma\sigma)|var(c)$. □

FINITE TREE IMPLEMENTATIONS

Using *tree–implementations*, we give a structure theorem for instantiation systems that strengthens previous similar results for composita [NERO59, Theorem 2.1], algebraic theories [GINA79, Theorem 2.8], and unification algebras [SCHM88, Theorem 5.14]. The presentation concludes with a comparison between *tree*-instantiation systems and order-sorted systems.

A *tree* instantiation system is one that is a subsystem of some TR#(A, V). A *natural* construct basis for a tree instantiation system is one where every construct is of the form $\langle a, v_1, \ldots, v_n\rangle$, with each v_i being a variable in the system (as well as in TR#(A, V)). If a tree system has a natural basis, then the set of all such terms of the form $\langle a, v_1, \ldots, v_n\rangle$ is the maximum natural construct basis. *Natural* construct bases for subsystems of first–order term instantiation are defined similarly.

Let C be a fixed set of nonvariables of $\mathfrak{B}$ such that every nonvariable is an instance of an element of C. We inductively define a *finite tree implementation* relation between trees in FTR#(C, V) and terms in B as follows: $\langle v\rangle$ is a tree implementation of v, for each v in V. If $c \in C$, $t = c\langle t_1, \ldots, t_n\rangle$, and for each $i = 1, \ldots, n$, x_i is a tree implementation of t_i, then $\langle c, x_1, \ldots, x_n\rangle$ is a tree implementation of t. In the remainder of this section, the symbols x, y, z, x_1, y_1, z_1, ... will vary over trees.

Proposition 6.3. Assume every nonvariable is an instance of a term in C.

A. A term has a finite tree implementation iff it is finitely generated.

B. If y is a finite tree implementation of s and of t, then s = t.

C. If terms are well-founded, then C is a construct basis.

Proof. Assertion A follows directly from the definitions involved. Assertion B is proved by induction on trees. The case where y = ⟨v⟩ is trivial. Suppose y = ⟨c, x_1, ..., x_n⟩, then by definition, there are terms s_1, ..., s_n such that x_i implements s_i, for each i, and s = c⟨s_1, ..., s_n⟩. Similarly, t = c⟨t_1, ..., t_n⟩, where x_i also implements t_i, for each i. By induction, $s_i = t_i$, for each i, so that s = t.

We prove Assertion C by contradiction. Consider a term t of minimal rank that is not finitely generated. Then t is neither a variable nor an element of C, but we may write t in the form t = c⟨t_1, ..., t_n⟩. Since each t_i has rank less than t, it is finitely generated, so that t itself is, contrary to choice of t. □

In view of Proposition 6.3B, we may define |x| to be the term that x implements, for each finite tree implementation x. A *type-preserving tree substitution* is a finite function φ from variables to finite tree implementations, with the property that v ≤ |φ(v)| and v ≠ |φ(v)|, for each v ∈ dom(φ).

For any instantiation system 𝔅 and construct basis C, let $𝔅^t$ be the structure defined as follows:

O^t is the set of all finite tree implementations of terms in O.

V^t is {⟨v⟩ | v ∈ V}.

S^t is the set of all type-preserving tree substitutions.

$*^t$ is the instantiation from Exercise 3.11 (straight plugging in).

Theorem 6.4.

A. The structure $𝔅^t$ is an instantiation system.

B. The function |_|: O^t -> O is a, type-preserving quotient homomorphism from $𝔅^t$ onto 𝔅.

C. |_| is one-to-one on variables and on constructs, with respect to the natural construct basis for $𝔅^t$.

D. |_| is one-to-one iff 𝔅 has unique quotients and unique constructs and C is skeletal.

Proof. Assertion A amounts to saying that O^t is instance-closed; this will be clear from the proof of Assertion B: For any type-preserving tree substitution φ, let |φ| = |φ(v_1)/v_1| + ... + |φ(v_n)/v_n|, where {v_1, ..., v_n} = dom(φ); we refer to |φ| as the *term* substitution that φ *implements*. The substitution |_|˜(φ) defined in the definition of a homomorphism clearly coincides with |φ|. We thus need to prove the identity |xφ| = |x||φ|, which we do by tree induction. The case x = ⟨v⟩ follows directly. Suppose x = ⟨c, x_1, ..., x_n⟩, then

$$\begin{aligned}
|x\phi| &= |\langle c, x_1\phi, \ldots, x_n\phi\rangle| && \text{– definition of tree substitution}\\
&= c\langle |x_1\phi|, \ldots, |x_n\phi|\rangle && \text{– definition of tree implementation}\\
&= c\langle |x_1||\phi|, \ldots, |x_n||\phi|\rangle && \text{– induction}\\
&= c\langle |x_1|, \ldots, |x_n|\rangle|\phi| && \text{– Proposition 3.10A}\\
&= |x||\phi| && \text{– definition of tree implementation.}
\end{aligned}$$

Thus, $|\_|$ is a homomorphism. It is full and type-preserving, because $v \leq x$ iff $v \leq |x|$. It is onto by Proposition 6.3A.

Assertion C follows directly from the definitions involved. Regarding Assertion D, if constructs are not unique or C is not skeletal, then $|\_|$ is clearly not one-to-one. Moreover, if $|\_|$ is one-to-one, we have unique quotients: Suppose $|x||\phi| = |x||\psi|$. Then, $x\phi = x\psi$, since $|\_|$ is one-to-one. Hence, $\phi|var(x) = \psi|var(x)$, since tree-systems have unique quotients. Hence, $|\phi||var(|x|) = |\psi||var(|x|)$. Finally, assume $\mathfrak{B}$ has unique-quotients and unique constructs and C is skeletal. We can prove $|\_|$ is one-to-one by tree induction: Assume $|x| = |y| = t$. If x is a variable, then $x = y = \langle t\rangle$, by definition of unique constructs. If $t = c\langle t_1, \ldots, t_n\rangle$, our assumptions imply that c is the label of both x and y, say $x = \langle c, x_1, \ldots, x_n\rangle$ and $y = \langle c, y_1, \ldots, y_n\rangle$. By unique quotients, $t_i = |x_i| = |y_i|$, for each i. Hence, $x_i = y_i$, by induction. Hence, $x = y$. □

Corollary 6.5. Every countable instantiation system is isomorphic to a quotient system of a subsystem of first-order term instantiation.

Proof. Any instantiation system has a basis, and first-order term instantiation is isomorphic with tree instantiation by Exercise 3.11B. This, together with Theorem 6.4, proves the Corollary. □

A construct basis *allows construct interpolations* iff whenever $v \leq t$ and $t \notin V$, there is a construct c such that $v \leq c \leq t$.

Exercise 6.6. Assume $\mathcal{A}$ is a subsystem of first-order term instantiation, and $\mathcal{A}$ has a natural construct basis C.

A. If $\mathcal{A}$ has restrictable terms, then the maximum natural construct basis for $\mathcal{A}$ allows construct interpolations.†

B. If $\mathcal{A}$ is an order-sorted system, then $\mathcal{A}$ is isomorphic to the quotient system of $\mathcal{A}^t$ induced by all equations of the form $\langle c, x_1, \ldots, x_n\rangle = \langle c', x_1, \ldots, x_n\rangle$, where $c = f(u_1, \ldots, u_n) \in T_1$, $c' = f(v_1, \ldots, v_n) \in T_2$, and $f(|x_1|, \ldots, |x_n|) \in T_1 \cap T_2$. Thus, as shown in Example 6.1, order-sorted systems can fail to have unique constructs.

C. If C allows construct interpolations, then $\mathcal{A}$ is order-sorted, with instance types for sorts, and with functions $f: T_1, \ldots, T_n \rightarrow T$, where $f(v_1, \ldots, v_n) \in C \cap T$, and, for each i, T_i, is the instance type of v_i.

† This is an extension of Exercise 5.6C. The assumption of restrictable terms cannot be weakened to weakly restrictable terms, by Example 5.2A, and this despite Exercise 3.12C.

TERM WEIGHTS

Term *weights* are defined relative to "label weights" for constructs. After listing basic properties, we investigate relationships among term weights, variable-occurrence counts, and label weights.

A *weight* function for a given instantiation system and construct basis is a function *wt* from terms to ordinals such that $wt(s) \leq wt(t)$, whenever $s \leq t$. A weight function is *additive* iff it is integer valued and, for some *label weight* function, lw, satisfies the identity $wt(c\langle t_1, \ldots, t_n\rangle) = lw(c) + wt(t_1) + \ldots + wt(t_n)$.

Exercise 6.7.

A. If $\mathcal{A}$ is a tree-instantiation system with a natural basis, then one may define an additive *tree*-weight function by taking wt(t) to be the number of nodes in t.

B. If terms in $\mathfrak{B}$ are well-founded, then ordinal rank is a weight function satisfying the condition $wt(c\langle t_1, \ldots, t_n\rangle) > wt(t_i)$, for each i.

C. If $\mathcal{A}$ has a weight function, then the terms of $\mathcal{A}$ are well-founded. (Consider an ill-founded term with minimal weight.)

D. An additive weight function is such that $wt(t\sigma) \leq wt(t\tau)$, provided $wt(v\sigma) \leq wt(v\tau)$, for all v.

E. For a given instantiation system and construct basis, there is at most one additive weight function, wt, satisfying the identities $wt(v) = wt(c) = 1$.

Exercise 6.8. Let $g: \mathcal{A} \twoheadrightarrow \mathfrak{B}$ be a surjective homomorphism that maps constructs to constructs.

A. Assume wt is an (additive) weight function for $\mathfrak{B}$; let wt_g be defined on terms of $\mathcal{A}$ by $wt_g(t) = wt(g(t))$; then wt is an (additive) weight function for $\mathcal{A}$.

B. Assume wt is an (additive) weight function for $\mathcal{A}$; assume $g(s) = g(t)$ implies $wt(s) = wt(t)$. Let $g^{\sim}(wt)$ be the function on terms of $\mathfrak{B}$ such that $g^{\sim}(wt)(g(t)) = wt(t)$, for all t in $\mathcal{A}$. Then $g^{\sim}(wt)$ is an (additive) weight function for $\mathcal{A}$.

The *variable occurrence counter*, *voc*(v, t), is given by the identities,

$$voc(u, u) = 1$$
$$voc(u, v) = 0, \text{ if } u \neq v$$
$$voc(v, c\langle t_1, \ldots, t_n\rangle) = voc(v, t_1) + \ldots + voc(v, t_n),$$

provided it is well-defined; in this case, we let $voc(E, t) = \Sigma\{voc(v, t) \mid v \in E\}$.

Proposition 6.9. Assume that wt is an additive weight function for $\mathcal{A}$. Assume, except in Assertion A, that voc is well-defined.

A. If for every v, there exists $r \geq v$ such that $wt(r) > wt(v)$, then voc is well-defined.

B. $wt(t(r/v)) = wt(t) + voc(v, t)\cdot(wt(r) - wt(v))$.

C. Let *cwt*(t) = 2·wt(t) − voc(var(t), t) − 1. Then cwt is the *construct weight*: cwt(v) = 0; cwt(c⟨t_1, ..., t_n⟩) = wt(c) + cwt(t_1) + ... + cwt(t_n).

D. Assume wt(v) = 1, for all v; extend lw by defining *lw*(t) = wt(t) − voc(var(t), t); this really is an extension of the label weight function, since lw(c⟨t_1, ..., t_n⟩) = lw(c) + lw(t_1) + ... + lw(t_n). Moreover, if v ∈ var(t), then lw(tσ) ≥ lw(t) + lw(vσ).

E. Suppose lw(c) = wt(v) = 1, for all c and v; if tσ is a construct and t is a nonvariable, then t is a construct, and σ|var(t) is a weakening for t.

Proof. We prove Assertions A and B together by induction, either assuming voc(v, t) is defined or using Assertion B to define it. We have, wt(v(r/v)) = wt(r) = wt(v) + voc(v, v)·(wt(r) − wt(v)). Similarly, if t = u and u ≢ v, then wt(u(r/v)) = wt(u) + voc(v, u)·(wt(r) − wt(v)). Finally, assume that for all terms of weight less than wt(t), voc(v, t) is defined and Assertion B holds. If t = c⟨t_1, ..., t_n⟩ is any construct factorization of t, we have,

wt(t(r/v))
 = lw(c) + wt(t_1(r/v)) + ... + wt(t_n(r/v))
 = lw(c) + wt(t_1) + ... + wt(t_n) + (voc(v, t_1) + ... + voc(v, t_n))·(wt(r) − wt(v)))
 = wt(t) + voc?(v, t)·(wt(r) − wt(v))).

The above identity is clearly independent of the chosen factorization for t. If wt(r) > wt(v), we also have voc?(v, t) = (wt(t(r/v)) − wt(t))/(wt(r) − wt(v)), showing that voc(v, t) is indeed well-defined.

Assertions C and D are left as exercises for the reader. For Assertion E, pick any v ∈ var(t), then 1 = lw(tσ) ≥ lw(t) + lw(vσ), by Assertion D. Since t is a nonvariable, it must have positive label weight, and so 0 ≥ lw(vσ). This shows that σ is a weakening for t, and that 1 ≥ lw(t), so that t is a construct. □

EXAMPLES

6.1. *Nonunique Constructs.* Consider the order-sorted instantiation system associated with the following type theory: $\mathcal{T}$ = {N, R}, N ⊆ R;

0, **1**, **i**, **j** ∈ N;
x, **y**, **i + x**, **x + i** ∈ R.

With respect to the construct basis {**0**, **1**, **i + x**, **x + 1**}, the term **1 + 1** has two different construct factorizations. This example has unique quotients, but not unique constructs.

6.2. *Constructs For Quotient Systems.* Starting with first-order term instantiation, consider the quotient system which is induced by the identity **P(S(x)) = x**. Then term occurrences are not well-founded, since [**S(x)**] = [**P(S(S(x)))**], so that [**S(x)**] occurs nontrivially in itself. In this example, the minimal nonvariables include [**P**(**x**)] but not [**S**(**x**)], since [**P**(**x**)] < [**S**(**x**)]. As a result, the term [**S**(**x**)] is not finitely generated over the set of minimal nonvariables.

6.3. *First-Order Schema Instantiation.* According to Corollary 6.5, alpha instantiation may be realized as a quotient system of a subsystem of first-order term instantiation. We now use such a realization to construct an extension of Example 2.3C whose constructs are similar to those normally associated with first-order logic.

A. Consider the order-sorted system given by the following typing axioms, where FUN and PRED are disjoint sets of function symbols containing infinitely many n-ary symbols, for each n. Define *TFS#* (for *term- and formula-schemas*), to be the order-sorted system axiomatized as follows: $\mathcal{T}$ = {BN, PT, TS, FS}; BN $\subseteq$ TS; PT $\subseteq$ TS;

pure terms:	$\mathbf{x}, \mathbf{x}_1, \mathbf{x}_2, \ldots \in$ PT $\cap$ V;
	f($\mathbf{x}_1, \ldots, \mathbf{x}_n$) $\in$ PT, for each n-ary f $\in$ FUN;
term schemas:	$\mathbf{z}, \mathbf{z}_1, \mathbf{z}_2, \ldots \in$ TS $\cap$ V;
	f($\mathbf{z}_1, \ldots, \mathbf{z}_n$) $\in$ TS, for each n-ary f $\in$ FUN.
bindable names:	$\mathsf{a}, \mathsf{b}, \mathsf{b}_1, \mathsf{b}_2, \ldots \in$ BN $\setminus$ V;
formula schemas:	$\boldsymbol{\alpha}, \boldsymbol{\beta}, \boldsymbol{\beta}_1, \boldsymbol{\beta}_2, \ldots \in$ FS $\cap$ V;
	$\neg\, \boldsymbol{\alpha}, \boldsymbol{\alpha} \rightarrow \boldsymbol{\beta} \in$ FS;
	($\exists$ a)($\boldsymbol{\alpha}$) $\in$ FS, (a unary construct) for each a $\in$ BN;
	p($\mathbf{z}_1, \ldots, \mathbf{z}_n$) $\in$ FS, for each n-ary p $\in$ PRED.

Notice that no bindable name is a variable or a pure term; individual constants (i. e., 0-ary function symbols) are pure terms, not bindable names.

B. A *term* is a term-schema that contains no term-schema variables; a *formula* is a formula schema that contains no term- or formula-schema variables. The terms and formulas determine a full subsystem, *TF#*, of TFS#. In fact, TF# is an ideal in TFS#.

C. Let *α-cnv* be the instantiation congruence on TF# induced by all equations of the form ($\exists$ a)(ψ(a/$\mathbf{z}$)) = ($\exists$ b)(ψ(b/$\mathbf{z}$)), where ψ(a/$\mathbf{z}$) is a formula and neither a nor b occurs in ψ. By abuse of notation, we also let *α-cnv* be the smallest containing congruence on TFS#. Notice that TF#/α-cnv is an ideal in TFS#/α-cnv, by Exercise 5.12A.

D. Notice that a term is a pure term iff it contains no bindable names. A name occurs *unbound* in a formula ϕ of TF# iff it occurs in every formula that is α-equivalent to ϕ. A *pure* formula is one in which no name occurs unbound. The pure terms and formulas form an ideal *PTF#* in TFS#, and determine a corresponding ideal PTF#/α-cnv in TFS#/α-cnv. The system PTF#/α-cnv is isomorphic to Example 2.3C.

E. Instantiation systems simpler than TFS#/α-cnv fail to give an adequate account of full first-order instantiation:

   i) If we allow bindable names to instantiate pure term variables, then $(\exists\ \mathbf{a})(\mathbf{a} < \mathbf{a})$ is an instance of $(\exists\ \mathbf{a})(\mathbf{a} < \mathbf{x})$.

   ii) If term schema variables are eliminated, then the construct $f(\mathbf{z}_1, \ldots, \mathbf{z}_n)$ has no sensible approximation in the smaller system.

   iii) If we allow alpha conversion for arbitrary formula schemas, then $(\exists\ \mathbf{b})(\mathbf{b} < \mathbf{a})$ is α-equivalent to $(\exists\ \mathbf{a})(\mathbf{a} < \mathbf{a})$.

F. In the system TFS#/α-cnv,

   i) substitutions are variable-preserving, by Exercise 5.2.
   ii) $t\sigma/t$ is defined, for any term t and schema substitution σ.
   iii) $\phi\sigma/\phi$ is defined, for any formula φ.
   iv) Every term and formula, considered as a schema, is rigid.

6.4 *Lambda Instantiation.* Consider the instantiation system associated with typed lambda calculus modulo α– and β–conversion. A type theory for this system may be given as follows: $\mathcal{T}$ is the smallest set of first-order terms that contains a given set $\mathcal{P}$ of "primitive" sort symbols and contains the term $T_1 \rightarrow T_2$, for each $T_1, T_2 \in \mathcal{T}$. Sorts are strict, in the sense that $T_1 \subseteq T_2$ implies $T_1 = T_2$. There are constant symbols of every type: $\mathbf{C}_P, \mathbf{D}_P, \ldots \in P$, for each primitive sort P; $\mathbf{F}_T, \mathbf{G}_T, \ldots \in T$, for each sort $T \in \mathcal{T} \setminus \mathcal{P}$. A term of the form $[\lambda\, v\, .\, t]$ is of type $T_1 \rightarrow T_2$, provided v is of type T_1 and t is of type T_2; $[s(t)]$ is of type T_2, provided s is of type $T_1 \rightarrow T_2$ and t is of type T_1. We write $s(t_1, \ldots, t_m)$ as an abbreviation for $s(t_1)\ldots(t_m)$, and write $\lambda\, v_1 . \ldots \lambda\, v_m\, .\, s$ as an abbreviation for $\lambda\, v_1 . (\ldots \lambda\, v_m\, .\, s)$.

A. We have the following:

   i) Every type is of the form $T_1 \rightarrow (\ldots \rightarrow (T_n \rightarrow P))$, with $P \in \mathcal{P}$.
   ii) Every term t may be written in the form $t = [\lambda v_1 . \ldots \lambda v_m . b(s_1(v_1, \ldots, v_m), \ldots, s_n(v_1, \ldots, v_m))]$, where no v_i occurs free in any s_j, and b is a variable or a constant symbol.
   iii) In the above form, m and the types of $v_1, \ldots, v_m$ are unique. Either b is unique or else $b = v_i$, for some unique i. Moreover, n and $[s_1], \ldots, [s_n]$ are unique.

B. As a result of Part A, the following is a set of constructs; it is the set suggested by Huet's unification algorithm [HUET75]:

$$\Pi_k\, \mathrm{Ss}(w_1, \ldots, w_n) = [\lambda v_1 . \ldots \lambda v_m . v_k(w_1(v_1, \ldots, v_m), \ldots, w_n(v_1, \ldots, v_m))],$$
$$\Lambda \mathrm{Ss}.F(w_1, \ldots, w_n) = [\lambda v_1 . \ldots \lambda v_m . F(w_1(v_1, \ldots, v_m), \ldots, w_n(v_1, \ldots, v_m))],$$
$$\Lambda \mathrm{Ss}.u(w_1, \ldots, w_n) = [\lambda v_1 . \ldots \lambda v_m . u(w_1(v_1, \ldots, v_m), \ldots, w_n(v_1, \ldots, v_m))],$$

where Ss is the sequence of types of $v_1, \ldots, v_m$, F is any constant symbol, $v_i \neq w_j$, $u \neq v_i$, $u \neq w_j$, for all i, j, and $1 \leq k \leq m$; in the third form, m and n must not both be 0. Notice that there are several useful special cases:

$F =$	$\Lambda \emptyset.F()$	— (m = n = 0)
$u(w_1, \ldots, w_n) =$	$\Lambda \emptyset.u(w_1, \ldots, w_n)$	— functional application (m = 0)
$\mathbf{K}_{Ss}.u =$	$\Lambda\ Ss.u(\,)$	— m-ary constant fn. (n = 0)
$\mathbf{\Pi}_k\ Ss =$	$\mathbf{\Pi}_k\ Ss(\,)$	— k^{th} m-ary projection fn. (n = 0)
$\mathbf{I}_S =$	$\mathbf{\Pi}_1\ S$	— identity fn. (m = 1, n = 0).

C. The nonrigid constructs are those of the third form, $\Lambda\ Ss.u(w_1, \ldots, w_n)$. For terms of the forms $\Lambda Ss.r(t_1, \ldots, t_n)$ or $\mathbf{\Pi}_k\ Ss(t_1, \ldots, t_n)$, we refer to Ss as an η-*signature*. Every term has a unique maximal η-signature. We write (S, Ss) for the sequence formed by prepending S to Ss. The consequences of β-reduction may be summarized as follows, where x is of type S:

i) $(\Lambda\ (S, Ss).u(w_1, \ldots, w_n))(x) = \Lambda\ Ss.u(w_1(x), \ldots, w_n(x))$.
ii) $(\mathbf{\Pi}_1\ (S, Ss)(w_1, \ldots, w_n))(x) = \Lambda\ Ss.x(w_1(x), \ldots, w_n(x))$.
iii) $(\mathbf{\Pi}_k\ (S, Ss)(w_1, \ldots, w_n))(x) = \mathbf{\Pi}_{k-1}\ Ss(w_1(x), \ldots, w_n(x))$, if k > 1.
iv) q = r, provided $q(v_1, \ldots, v_m) = r(v_1, \ldots, v_m)$, q and r both have η-signature $Ss = S_1, \ldots, S_m$, and $v_i \in S_i$, for each i.

D. Instantiation in typed lambda calculus is isomorphic to the quotient system $\mathcal{L}^t/E$, where $\mathcal{L}^t$ is the tree-system associated with the construct basis given in Part B and E is the congruence induced by the identities of Part C.

6.5. *Multiple-Arity Operators.*

A. We begin with one multiple-arity function symbol **S**, and all terms of the form $\mathbf{S}(t_1, \ldots, t_n)$, for n > 0. Instantiation is by straight plugging in. We include one construct of the form $\mathbf{S}(\mathbf{x}_1, \ldots, \mathbf{x}_n)$, for each n.

B. To the above example, add *list* variables $\mathbf{xs}, \mathbf{xs}_1, \mathbf{xs}_2, \ldots$, and terms of the forms $\langle t_1, \ldots, t_n \rangle$ and $\mathbf{S}(vs)$, where vs is any list variable. Substitutions replace list variables with lists or other list variables. We obtain a new set of constructs by replacing constructs of the form $\mathbf{S}(\mathbf{x}_1, \ldots, \mathbf{x}_n)$ with the single new construct $\mathbf{S}(\mathbf{xs})$ and adding infinitely many constructs of the form $\langle \mathbf{x}_1, \ldots, \mathbf{x}_n \rangle$.

C. To the above example, add terms of the forms $\mathbf{S}(t_1, \ldots, t_n \mid vs)$ and $\langle t_1, \ldots, t_n \mid vs \rangle$. Terms of this latter form are considered to be list-valued. If σ is a substitution of the form $(s_1, \ldots, s_n)/vs$, then $\langle t_1, \ldots, t_n \mid vs \rangle\sigma$ is the list $\langle t_1\sigma, \ldots, t_n\sigma, s_1, \ldots, s_n \rangle$. For this system, we need just two constructs, namely $\mathbf{S}(vs)$ and $\langle v \mid vs \rangle$.

6.6. *Variable–Arity Tree Implementations.* A binary construct c is *associative* iff it satisfies the identity c⟨r, c⟨s, t⟩⟩ = c⟨c⟨r, s⟩, t⟩. The concatenation construct from Example 2.2A is associative in this sense.

A. Tree implementations exist for associative constructs, but they are not unique. However, if we choose to implement c⟨r, c⟨s, t⟩⟩ as a ternary construct by using a tree of the form ⟨c, x_r, x_s, x_t⟩, then we again have unique tree implementations for terms, and a given tree label can have multiple arities.

B. This suggests a new set of constructs for the original system, consisting of all terms of the form **S**(x_1, ..., x_n) = c⟨x_1, c⟨x_2, ... x_n⟩...⟩, as in Example 6.5A.

The operator **S** of Example 6.5A does not have multiple arity in a meaningful sense, because, for example, **S**(x_1) and **S**(x_1, x_2) are different constructs. Correspondingly, there is no way, in this example, to reason about applying **S** to an arbitrary list. In Example 6.5B, we may now reason about all instances of the construct **S**(**x**s), but there are still infinitely many *list* constructs of the form ⟨x_1, ..., x_n⟩, and we have merely displaced the original problem. In Example 6.5C, we have gained a general ability to reason about multiple–arity operators by recasting them as binary operators; this approach to reasoning about lists is essentially the one taken by Prolog [CLOC81]. Finally, Example 6.6 suggests the origin of the construct-schemas considered in Example 6.5A.

SECTION 7
UNIFICATION – AN ALGORITHM AND ITS SOUNDNESS

The major goal of Sections 7 through 9 is to produce a broadly applicable algorithm that is sound, complete, and efficient. An additional pragmatic goal is to facilitate human use by avoiding capricious variable renamings. This latter goal has a relatively subtle impact on the basic algorithm, but significantly affects proofs of soundness and completeness, and also uncovers an interesting opportunity for optimization. A secondary goal is to give a clear, rigorous development by taking advantage of abstraction and modularity. In particular, we wish to correlate soundness and completeness for unification in abstract instantiation systems with analogous requirements in concrete term implementations. Another secondary goal is to provide a simple, well-defined interface for the algorithm, in order to encourage use in a variety of software reasoning systems.

A basic understanding of unification techniques is readily achieved through the use of simple, abstract transformations of the sort introduced by Gallier and Snyder [GALL88], as can be seen in the work of Schmidt-Schauss and Siekmann [SCHM88, Sections 6 & 9]. The transformations we will use are somewhat more complex, but support prudent use of variable renamings.†

We consider nondeterministic unification algorithms in order to avoid tangential issues such as backtracking versus parallelism. The approach leans heavily on ideas of Martelli and Montanari [MART82]. It includes ideas from Algorithm 2 of Martelli, Moiso, and Rossi [MART86], but makes no explicit use of term rewriting.

The main algorithm to be studied can be understood abstractly and shown to be sound in any *effectively supported* instantiation system. In this section, we specify the problem to be solved, introduce some basic strategies, present a solution, prove its soundness, and consider its use with various instantiation systems. Section 8 shows that a variant of the algorithm is complete, if terms are acyclic and weakly restrictable and are appropriately implemented. Low-level refinements needed to guarantee efficiency are presented in Section 9, along with a computational complexity analysis.

REQUIREMENTS

We begin by generalizing the unification problem introduced in Section 2 to that of finding unifiers for *multiequations* and *term systems*. To facilitate a precise treatment of soundness and completeness, we use a model of computation in

† While the "unfolding" and "partial solution" rules in [SCHM88] are quite general, they rely heavily on variable renamings.

which algorithms define binary *I/O* relations. Finally, the specification of low-level *support* routines provides a basis for computing unifiers in a wide variety of instantiation systems. It is intended that the four main support routines be table-driven and derived from an evolving mathematical theory as follows:

sub_both(T, T_1, T_2): Theorems of the forms $T \subseteq T_1 \cap T_2$ and $T \subseteq T_1$.
restricts(σ, c, T): Theorems of the form $c\langle t_1, ..., t_n\rangle \in T$ (i. e., $c\sigma \in T$).
unifies_2c(σ, c, d): Exercise 7.1B and theorems of the form $c\langle s_1, ..., s_m\rangle = d\langle t_1, ..., t_n\rangle$ (i. e., $c\sigma = d\sigma$).
factors(c, σ, t): Construct formation, as defined by the abstract syntax of the theory: $t = c\langle t_1, ..., t_n\rangle = c\sigma$.

The support procedures are specified in an Ada-like notation [DOD83] that, for convenience, relies on typing conventions already given in the text.

A *multiequation* is a finite multiset of terms in which no variable occurs more than once. The variables D, E, F, G, H, ... will range over multiequations, unless otherwise specified. We will freely use set notation for multisets. Thus, $s \in E$ and $t \in E \setminus \{s\}$ allows the possibility that $s = t$.† Define $E\sigma$ to be $\{t\sigma \mid t \in E\}$. Define $t_1 \approx ... \approx t_n$ to be the largest multiequation contained in the multiset $\{t_1, ..., t_n\}$. Thus, $v \approx v$ is $\{v\}$, not the multiset $\{v, v\}$. The *(multiequation) union* of two multiequations, written $E \uplus F$, is just the multiset $E \cup F$ with duplicate variables removed.

A *term system* is a finite multiset of non-empty multiequations. The variables $\mathfrak{D}$, $\mathfrak{E}$, $\mathfrak{F}$, $\mathfrak{G}$, ... will range over term systems. Define $\mathfrak{E}\sigma$ to be $\{E\sigma \mid E \in \mathfrak{E}\}$. Define $\sigma_1 \approx ... \approx \sigma_n$ to be the term system $\{v\sigma_1 \approx ... \approx v\sigma_n \mid v \in \mathrm{dom}(\sigma_i)$, for some i$\}$. Extend the var operator to multiequations and term systems by taking unions in the obvious way.

A substitution σ is a *unifier* for a multiequation $t_1 \approx ... \approx t_n$ iff $t_1\sigma = ... = t_n\sigma$. We say σ is a *(simultaneous) unifier* for a term system $\mathfrak{E}$ iff σ unifies each multiequation in $\mathfrak{E}$. Notice that τ is a unifier for the term system $\sigma_1 \approx ... \approx \sigma_n$ iff $\sigma_1\tau = ... = \sigma_n\tau$ [cf VANV75].

Let U, W, Y, ... vary over finite subsets of V. Define $\sigma \leq \tau$ [U] iff $(\sigma\eta)|U = \tau|U$, for some η. In particular, we write $\sigma \leq \tau$ iff $\sigma \leq \tau$ $[\mathrm{dom}(\sigma) \cup \mathrm{dom}(\tau)]$. If σ is a unifier of a term system $\mathfrak{E}$, then σ is a *more general* unifier than τ iff $\sigma \leq \tau$ $[\mathrm{var}(\mathfrak{E})]$. Finally, a set Σ of unifiers for $\mathfrak{E}$ is *complete* iff every unifier of $\mathfrak{E}$ is less general than some element of Σ.

† The few steps where the distinction between sets and multisets matters will be explicitly called c
The need for multisets derives from the lack of a unit-time equality test for arbitrary terms.

Exercise 7.1.

A. If σ unifies $\{s_1 \approx t_1; \ldots; s_n \approx t_n\}$, then σ unifies $c\langle s_1, \ldots, s_n\rangle \approx c\langle t_1, \ldots, t_n\rangle$.

B. If η is a renaming from var(d) to var(c), and dom(η) $\cap$ var(c) = $\emptyset$, then η unifies $c \approx d$.

C. If σ unifies $\mathcal{S}$ and $\sigma \leq \tau$ [dom(σ)], then τ unifies $\mathcal{S}$.

D. If $\sigma \leq \tau$ [dom(σ)] and cdm(σ) $\cap$ U = $\emptyset$, then $\sigma \leq \tau$ [U].

E. Assume d is a rigid construct and var(c) $\cap$ var(d) = $\emptyset$. Then $\{\gamma \mid c\gamma = d$ and dom(γ) $\subseteq$ var(c)$\}$ is a complete set of unifiers for $c \approx d$.

F. Assume terms are restrictable and var(c) $\cap$ $\{v_1, \ldots, v_n\} = \emptyset$. Then the unifiers η of $v_1 \approx \ldots \approx v_n \approx c$ such that $\eta|$var(c) is a weakening for c form a complete set.

G. The *equivalencing problem* [cf TARJ75, AHO85] is to devise a procedure that, given a finite relation $R = \{\langle u_1, v_1\rangle, \ldots, \langle u_n, v_n\rangle\}$, returns the smallest equivalence relation E containing R. Assume, for simplicity, that the u_i and v_i are first-order variables. If σ is a most-general unifier for the equation $\mathsf{f}(u_1, \ldots, u_n) \approx \mathsf{f}(v_1, \ldots, v_n)$, then u E v iff $u\sigma = v\sigma$, for all u, v in the field of R. Thus, any unification procedure for first-order terms solves the equivalencing problem.

We assume that every procedure call has a *return-code* as an implicit output. The return-code is either an *exception* or one of two non-exceptions, *success* and *nontermination*. A particular execution of a procedure is successful iff its return code is success. A procedure is *deterministic* iff it has a single-valued I/O relation. A *sound unification* procedure accepts a term system $\mathcal{S}$ as input and, if successful, returns a substitution σ that unifies $\mathcal{S}$.

Effective support for an instantiation system $\mathfrak{B}$ consists of a construct basis with the property that every renaming of a construct is a construct, together with implementable functions and procedures, as specified in the generic package unification$_1$ given below. We assume that the indicated support procedures either succeed as indicated in their exit conditions or else fail by raising a *clash_failure* when success is not possible. The unify procedure itself either returns a unifier, fails to terminate, or else raises a *clash_failure* or *cycle_failure* exception.

Several assumptions are readily satisfied by the support procedures, even though they are not assumed in the proofs. If the nonempty intersection of two types is always a type, one could assume that sub_both(T, T_1, T_2) succeeds iff $T_1 \cap T_2 \neq \emptyset$, in which case, $T = T_1 \cap T_2$. If terms are restrictable, one could assume that the substitutions returned by restricts(σ, c, T) are weakenings (by Exercise 7.1F). If c is rigid and σ is returned by unifies_2c(σ, c, d), one could assume that $\sigma|var(c) = \epsilon$ (by Exercise 7.1E). If t is a construct, one could assume, after factors(c, σ, t), that c = t and σ = ε. These assumptions are exploited in the presented examples.

```
Generic package unification1
   type O, V, S, C;
   with  tσ, στ, σ + τ, σ|U, dom(σ), ρᵛ, t/v, ε, var(c),
            t ∈ V, t ∈ C, u = v, c = d, T_u = T_v, T_u ⊆ T_v;
            -- operations are given relative to the instantiation
            -- system (O, V, S, *) with construct basis C;
         procedure sub_both(T: out; T1, T2: in);
            -- exit T ⊆ T1 ∩ T2;
         procedure restricts(σ: out; c: in; T: in);
            -- exit cσ ∈ T and dom(σ) ⊆ var(c);
         procedure unifies_2c(σ: out; c, d: in);
            -- entry var(c) ∩ var(d) = ∅;
            -- exit σ unifies c ≈ d and dom(σ) ⊆ var(c) ∪ var(d).
         procedure factors(c, σ: out; t: in);
            -- entry t ∉ V;
            -- exit t = cσ and dom(σ) ⊆ var(c).
   end with;
is
   procedure unify(σ: out; 𝔇: in);
      -- exit σ unifies 𝔇;
   clash_failure, cycle_failure: exception;
end unification1;
```

BASIC STRATEGIES

In attempting to unify a term system 𝔖, we use the following basic strategies:

- Identify any multiequations in the *base* of a *triangular* enumeration of 𝔖.
- Perform a *reduction*: fragment a base multiequation into a simpler term system and, possibly, identify a portion of an emerging partial unifier δ.
- *Merge* overlapping fragments into a single multiequation.

The following paragraphs discuss base multiequations and merging.

A multiequation E is *isolated* in 𝔖 iff $E \in 𝔖$ and $E \cap F \cap V = \emptyset$, whenever $F \in 𝔖 \setminus \{E\}$. We say 𝔖 is *merged* iff every element of 𝔖 is isolated. Thus, 𝔖 is merged iff $E \uplus F = E \cup F$, for all $E \in 𝔖$ and $F \in 𝔖 \setminus \{E\}$. Define 𝔉 to be *connected* iff 𝔉 is not of the form $𝔊 \cup ℌ$, where $𝔊 \neq \emptyset$, $ℌ \neq \emptyset$, and $G \cap H \cap V = \emptyset$, for all $G \in 𝔊$ and $H \in ℌ$. A *component* of 𝔖 is a maximal connected submultiset of 𝔖. Define *merge*(𝔖) to be $\{\biguplus 𝔉 \mid 𝔉$ is a component of $𝔖\}$.

We say E *precedes* F iff $E \cap \mathrm{var}(F \setminus V) = \emptyset$. Notice that if $E \cap V = \emptyset$, then E precedes F, for any F. We define $base_{\mathcal{E}}$ to be the set of all isolated E in $\mathcal{E}$ such that E precedes every element of $\mathcal{E}$. If $\mathcal{E}$ is a merged system, then an enumeration $E_1, \ldots, E_n$ of $\mathcal{E}$ is *triangular* iff E_i precedes E_j, whenever $i \leq j$. In this case, E can be the first element of a triangular enumeration of $\mathcal{E}$ iff $E \in base_{\mathcal{E}}$ iff $E \cap \mathrm{var}((\mathcal{E} \setminus \{E\}) \cup \{E \setminus V\}) = \emptyset$.

If $\epsilon \approx \sigma$ is merged, then σ has a triangular enumeration iff $\epsilon \approx \sigma$ does. Regarding sharpness of Proposition 7.2, non-triangular term systems can be unifiable in systems where there are term-occurrence cycles. We have already seen substitutions σ that have fixed points without having triangular enumerations. (Examples 2.5 and 4.6 both allow this.) If τ is a fixed point for the nontriangular substitution σ, and if $\epsilon \approx \sigma$ is merged, then τ unifies the nontriangular system $\epsilon \approx \sigma$.

Proposition 7.2. If terms are acyclic and $\mathcal{E}$ is a unifiable merged system, then $\mathcal{E}$ has a triangular enumeration [cf BURC87, Prop. 3.2; MART82, Thm. 3.3].

Proof. Suppose $\mathcal{E}$ fails to have a triangular enumeration, with $\mathcal{E}$ as small as possible. Let τ be a unifier of $\mathcal{E}$. There is no element of $\mathcal{E}$ that precedes every element of $\mathcal{E}$, since otherwise, $\mathcal{E}$ would have a triangular enumeration with first element E, since $\mathcal{E} \setminus \{E\}$ has a triangular enumeration by minimality of $\mathcal{E}$. Choose $E_1, \ldots, E_n \in \mathcal{E}$ so that each E_i fails to precede E_{i+1}, and $E_n = E_1$. For each $i < n$, we may choose $v_i \in E_i$ and $t_{i+1} \in E_{i+1} \setminus V$ so that $v_i \in \mathrm{var}(t_{i+1})$. Then, for each $i < n$, $t_i\tau$ properly occurs in $t_{i+1}\tau$, since $v_i\tau = t_i\tau$ and $v_i \in \mathrm{var}(t_{i+1})$. Similarly, $v_1\tau = t_n\tau$, so that $t_n\tau$ properly occurs in $t_2\tau$. This contradicts the fact that terms are acyclic. □

Exercise 7.3.

A. Every element of $\mathcal{E}$ belongs to a unique component of $\mathcal{E}$, and merge($\mathcal{E}$) is merged.

B. $\mathcal{E}$ is a merged system iff $\mathcal{E} = \mathrm{merge}(\mathcal{E})$.

C. $\mathrm{var}(\mathrm{merge}(\mathcal{E})) = \mathrm{var}(\mathcal{E})$.

D. If $E \cap V = \emptyset$, then $E \in \mathrm{merge}(\mathcal{E})$ iff $E \in \mathcal{E}$ iff $E \in base_{\mathrm{merge}(\mathcal{E})}$.

E. If E is isolated in $\mathcal{E}$, then $E \in base_{\mathcal{E}}$ iff $E \in base_{\mathrm{merge}(\mathcal{E})}$.

F. $\mathrm{merge}(\mathcal{E} \cup \mathrm{merge}(\mathcal{F})) = \mathrm{merge}(\mathcal{E} \cup \mathcal{F})$.

G. σ unifies merge($\mathcal{E}$) iff σ unifies $\mathcal{E}$; in this case, $\mathrm{merge}(\mathcal{E})\sigma$ contains the same terms as $\mathcal{E}\sigma$.

H. Assume $\mathcal{E}$ is a merged system. If $\mathrm{merge}(\mathcal{E} \cup \{F\}) = \mathcal{F}$ and $G \in \mathcal{E}$, then $G \in base_{\mathcal{F}}$ iff $G \in base_{\mathcal{E}}$ and either $G \cap \mathrm{var}(F) = \emptyset$ or $F \subseteq G \cap V$.

ALGORITHMIC SPECIFICATION

The algorithm iteratively reduces a term system $\mathcal{E}$ and transfers information into a partial unifier δ. On each iteration of the unify loop, an optional merge ensures the availability of a base multiequation E that is reduced, simplifying E and enlarging $\mathcal{S}$ and/or δ. A given reduction can act on two variables, a variable and a nonvariable, or two nonvariables, using the support procedures sub_both, restricts, and unifies_2c, respectively. In the latter two cases, some renaming of variables is necessary. The specification allows all new variables to be chosen, but such a choice makes execution traces harder to follow and could result in uglier unifiers. In the specification, the key word "any" is used to denote an arbitrary choice in order to suppress low-level design details.

Package body *unification₁* is

```
procedure unify(δ: out := ε; 𝔇: in)
is 𝒮 := 𝔇; E; U;
begin
   pick any U with U ⊇ var(𝔇);
   loop exit when 𝒮 = ∅;
      select any
         base_𝒮 ≠ ∅      => null;
         true            => 𝒮 := merge(𝒮);
                            if base_𝒮 = ∅ then raise cycle_failure; end if;
      end select;
      pick any E ∈ base_𝒮;
      select any
         size(E ∩ V) > 1                 => vv_reduce(δ, E, 𝒮, U);
         size(E \ V) > 1                 => cc_reduce(E, 𝒮, U);
         E ∩ V ≠ ∅ and E \ V ≠ ∅         => vc_reduce(δ, E, 𝒮, U);
         size(E) = 1   => 𝒮 := 𝒮 \ {E};
      end select;
   end loop;
end unify;

procedure vv_reduce(δ, E, 𝒮, U: in out)
is u, v, w; γ;
begin
   pick u ∈ E; pick v ∈ E \ {u};
   sub_both(T, T_u, T_v);
   pick any w ∈ {v, u} ∪ (V \ U) so that T_w = T; U := U ∪ {w};
   γ := w/u + w/v; δ := δγ;
   𝒮 := 𝒮 \ {E}; E := {w} ∪ (E \ {u, v}); 𝒮 := {E} ∪ 𝒮;
end vv_reduce;
```

```
    procedure cc_reduce(E, ℰ, U: in out)
    is s, t; c, d; α, β, γ;
    begin
       pick s ∈ E \ V; pick t ∈ E \ (V ∪ {s});
       get_factor(c, α, s, ∅, U); get_factor(d, β, t, var(c), U);
       unifies_2c(γ, c, d); rename_cdm(γ, var(c ≈ d), U);
       ℰ := ℰ \ {E}; E := {cγ} ∪ (E \ {s, t});
       ℰ := ℰ ∪ {E} ∪ (γ ≈ α + β);
    end cc_reduce;
    procedure vc_reduce(δ, E, ℰ, U: in out)
    is v; t; c; α, γ, ψ;
    begin
       pick v ∈ E; pick t ∈ E \ V;
       get_factor(c, α, t, ∅, U);
       restricts(ψ, c, T_V); rename_cdm(ψ, var(c), U);
       γ := cψ/v + ψ;
       δ := δ(γ|{v});
       ℰ := ℰ \ {E}; E := {cγ} ∪ (E \ {v, t});
       ℰ := {E} ∪ ℰ ∪ (γ|var(c) ≈ α);
    end vc_reduce;
    procedure get_factor(c, α: out; t, W: in; U: in out)
    is c; σ; η; μ;
    begin
       factors(d, σ, t);
       pick any renaming μ such that
          dom(μ) ⊇ dom(σ) ∪ (W ∩ var(d)) and cdm(μ) ∩ U = ∅;
       c := dμ; α := (μ^v σ)|var(d); U := U ∪ var(c);
    end get_factor;
    procedure rename_cdm(γ: in out; W: in; U: in out)
    is η;
    begin
       pick any renaming η such that
          dom(η) ⊇ cdm(γ) \ (W \ dom(γ)) and cdm(η) ∩ U = ∅;
       γ := (γη)|dom(γ); U := U ∪ cdm(η);
    end rename_cdm;
end unification₁;
```

Use of the unify algorithm is illustrated in Examples 7.1 through 7.4. It avoids problems of the sort illustrated by Examples 7.5 and 7.6. Interestingly, Algorithm 3 of Martelli and Montanari [MART82] also fortuitously avoids these problems, even though they do not arise for the special case of first-order term unification. Jaffar's simpler algorithm [JAFF84], by way of contrast, is less robust. While the unify algorithm avoids some obvious pitfalls, it can fail to terminate. Both subsystems and quotient-systems of first-order term instantiation can allow nontermination (see Examples 7.2 and 7.4).

The design of the algorithm was based on the following observations: Early reduction facilitates detection of nonunifiable term systems. Early merging can reduce nondeterminacy by eliminating unsuccessful branches (see Example 7.5). Early processing of base multiequations reduces the amount of retyping needed (see Example 7.6). The variable renaming strategies used in the algorithm were arrived at by simply looking at what is needed to make the proofs work.

SOUNDNESS OF THE ALGORITHM

The proof of soundness uses traditional software verification methodology [cf CRAI87, GOOD88]. For exit conditions, we use a prime sign (′) to mark the output value of an in-out parameter. Loop-invariants either hold at the top of a loop or else relate values at the top to values at the bottom, in which case a prime sign marks the corresponding value of a variable at the bottom of the loop.

The formal entry and exit conditions used in the proof are stated as part of the soundness theorem, because they are needed for later results.

Theorem 7.4 (soundness). Assume an effectively supported instantiation system in which substitutions are variable-preserving. Then the unify procedure is sound. The formal entry and exit conditions in Table 7.1 hold after successful execution of the indicated procedures. Moreover, only those exit conditions involving unification depend on assumptions regarding unification.

Proof. To prove soundness, suppose σ is returned by unify(σ, $\mathfrak{D}$). Then, σ is the value of δ at the end of the unify loop; we can easily prove σ unifies $\mathfrak{D}$, assuming the following loop invariants:

unify loop inv: for some ρ, $\sigma = \delta\rho$ and ρ unifies $\mathfrak{S}$; (i)

$\mathrm{var}(\mathfrak{S}) \cup \mathrm{dom}(\delta) \cup \mathrm{cdm}(\delta) \subseteq U \subseteq U'$; (ii)

Applying invariant (i) on entry to the loop, we see that σ is of the form $\sigma = \delta\rho$, where $\delta = \epsilon$ and ρ unifies $\mathfrak{S}$, so that σ unifies $\mathfrak{D}$.

The second loop invariant is easily proved by loop induction, so we prove only the first, by reverse loop induction, relying on the lower-level exit conditions in Table 7.1. The first invariant holds trivially on termination, with $\rho = \epsilon$, since $\mathfrak{S} = \emptyset$. For the induction step, we need to show that the invariant remains true before an arbitrary reduction preceded by an optional merge. Merging preserves the invariant by Exercise 7.3G, so we may concentrate on reduction.

unify exit:	δ unifies $\mathfrak{D}$;	
vv_reduce entry:	$size(E \cap V) > 1$; $var(E) \subseteq U$; $E \cap var(E \setminus V) = \emptyset$;	
vv_reduce exit:	exists $u \in E$, $v \in E \setminus \{u\}$, $w \in \{v, u\} \cup (V \setminus U)$, γ such that,	
	$T_w \subseteq T_u \cap T_v$, $\gamma = w/u + w/v$, $\delta' = \delta\gamma$,	(i)
	$U' \supseteq U \cup \{w\}$,	(ii)
	$E' = E \setminus \{u, v\} \cup \{w\}$, $\mathcal{E}' = \mathcal{E} \setminus \{E\} \cup \{E'\}$;	(iii)
cc_reduce entry:	$size(E \setminus V) > 1$; $var(E) \subseteq U$; $E \cap var(E \setminus V) = \emptyset$;	
cc_reduce exit:	exists $s \in E$, $t \in E \setminus \{s\}$, c, d, α, β, γ, U_1, such that	
	$s = c\alpha$, $t = d\beta$, γ unifies $c \approx d$,	(i)
	$U' \supseteq U_1 \cup var(\gamma) \cup var(d)$, $U_1 \supseteq U \cup var(c)$,	(ii)
	$E' = E \setminus \{s, t\} \cup \{c\gamma\}$,	
	$\mathcal{E}' = \mathcal{E} \setminus \{E\} \cup \{E'\} \cup (\gamma \approx \alpha + \beta)$,	(iii)
	$var(c) \cap U \subseteq var(s)$, $var(d) \cap U_1 \subseteq var(t)$,	
	$var(c) \cap var(d) = \emptyset$†,	(iv)
	$dom(\alpha) = var(c) \setminus U$, $dom(\beta) = var(d) \setminus U_1$,	(v)
	$dom(\gamma) \subseteq var(c \approx d)$,	
	$cdm(\gamma) \cap U \subseteq var(c \approx d) \setminus dom(\gamma)$;	(vi)
vc_reduce entry:	$E \cap V \neq \emptyset$; $E \setminus V \neq \emptyset$; $var(E) \subseteq U$; $E \cap var(E \setminus V) = \emptyset$;	
vc_reduce exit:	exists $v \in E$, $t \in E \setminus V$, c, α, γ such that	
	$t = c\alpha$, γ unifies $v \approx c$, $\delta' = \delta(\gamma \vert \{v\})$,	(i)
	$U' \supseteq U \cup var(c\gamma) \cup var(\gamma \vert var(c) \approx \alpha)$,	(ii)
	$E' = E \setminus \{s, v\} \cup \{c\gamma\}$,	
	$\mathcal{E}' = \mathcal{E} \setminus \{E\} \cup \{E'\} \cup (\gamma \vert var(c) \approx \alpha)$,	(iii)
	$var(c) \cap U \subseteq var(t)$,	(iv)
	$dom(\alpha) = var(c) \setminus U$,	(v)
	$dom(\gamma \vert var(c)) \subseteq var(c)$,	
	$cdm(\gamma) \cap U \subseteq var(c) \setminus dom(\gamma)$;	(vi)
get_factor entry:	$t \notin V$; $var(t) \subseteq U$;	
get_factor exit:	$t = c\alpha$;	(i)
	$U' \supseteq U \cup var(c)$;	(ii)
	$var(c) \cap U \subseteq var(t) \setminus W$;	(iii)
	$dom(\alpha) \subseteq var(c) \setminus U$	(iv)
rename_cdm exit:	for some renaming η,	
	$\gamma' = (\gamma\eta) \vert dom(\gamma)$,	(i)
	$U' \supseteq U \cup cdm(\eta)$, $cdm(\eta) \subseteq V \setminus U$,	(ii)
	$cdm(\gamma') \cap U \subseteq W \setminus dom(\gamma)$;	(iii)

Table 7.1: Entry and Exit Conditions for unification$_1$

† Disjointness of var(c) and var(d) is used only as an entry condition for unifies_2c.

Some $E \in base_{\mathcal{E}}$ has been picked, and comparing values before and after the reduction, we have,

$\mathcal{E}' = \mathcal{E} \setminus \{E\} \cup \mathcal{F}$, for some $\mathcal{F}$;
$\sigma = \delta'\rho' = \delta\theta\rho'$, where θ is γ, ϵ, or $\gamma|var(c)$, depending on the reduction;
ρ' unifies $\mathcal{E}'$, by induction.

Taking $\rho = \theta\rho'$, we have ρ unifies $\mathcal{E} \setminus \{E\}$, since $dom(\theta) \cap var(\mathcal{E} \setminus \{E\}) = \emptyset$, since $E \in base_{\mathcal{E}}$ and $dom(\theta) \subseteq E \cap V$, in all three cases. Thus, it suffices to show ρ unifies E.

Suppose vv_reduce was picked, and u, v were selected from E. Then, ρ unifies $u \approx v$, since θ does. If $t \in E \setminus \{u, v\}$, then ρ unifies $v \approx t$, since $u\rho = u\theta\rho' = w\rho' = t\rho' = t\theta\rho'$, using choice of θ (vv_reduce exit (i)), ρ' unifies E', and $u, v \notin var(t)$ (since $E \cap var(E \setminus V) = \emptyset$, since $E \in base_{\mathcal{E}}$). Hence, ρ unifies E.

Next, suppose cc_reduce was selected, so that $\delta' = \delta$, $\rho' = \rho$, and s, t have been selected from $E \setminus V$. From cc_reduce exit conditions (ii) and (iv), $var(c) \subseteq U_1$ and $dom(\beta) \subseteq V \setminus U_1$, so that $var(c) \cap dom(\beta) = \emptyset$ (where U_1 is the value of U after the first call to get_factor). Similarly, by exit conditions (iv) and (v), $var(d) \cap U_1 \subseteq U$ and $dom(\alpha) \subseteq U_1 \setminus U$, so that $var(d) \cap dom(\alpha) = \emptyset$. We have, $s\rho = c\alpha\rho = c(\alpha + \beta)\rho = c\gamma\rho = d\gamma\rho = d(\alpha + \beta)\rho = d\beta\rho = t\rho$, using the above observations, ρ unifies $\gamma \approx \alpha + \beta$, and γ unifies $c \approx d$. If $r \in E \cap E'$, then $r\rho = c\gamma\rho = t\rho$, using $\rho = \rho'$, ρ' unifies E', and the previous step. Hence, ρ unifies E.

Finally, suppose vc_reduce was selected. Then $\delta' = \delta\theta$, $\theta = c\gamma/v$, and $\rho = \theta\rho'$. We have, $t\rho = c\alpha\rho = c\gamma\rho = v\gamma\rho = v\theta\theta\rho' = v\theta\rho' = v\rho$, using ρ unifies $\alpha \approx \gamma|var(c)$, γ unifies $c \approx v$, and $cdm(\theta) \cap dom(\theta) = \emptyset$. If $s \in E \cap E'$, then $s\rho = s\theta\rho' = s\rho' = t\rho' = t\rho$, using $v \notin var(s)$, ρ' unifies E', and $v \notin var(t)$. Hence, ρ unifies E.

We omit the straightforward verification of the lower-level procedures, namely vv_reduce, cc_reduce, vc_reduce, get_factor, and rename_cdm. □

CHOICE OF INSTANTIATION SYSTEM

The following observations discuss the utility of the unify algorithm and are not essential to later sections.

The restriction to systems with variable-preserving substitutions will be overcome in Section 8. For the unify algorithm to be complete, terms must be acyclic, because of our reliance on triangular form. This restriction also applies to Algorithm 2 of Martelli et al. [MART86]. Terms must be weakly restrictable in order to successfully unify arbitrary pairs of variables, but this latter restriction seems to be fairly mild, in view of Proposition 5.3B, Proposition 5.7, and Theorem 6.4.

Non-removable inefficiencies occur in vc_reduce when $c\gamma$ is not a construct, and in cc_reduce when $d\gamma$ is not a construct. If variables and constructs are restrictable and this fact is reflected in the support procedures, then all terms are restrictable, as a result of Exercise 7.5, so that inefficiencies need not occur in vc_reduce in this case. If constructs are rigid, then we may assume $c\gamma$ is always a construct in cc_reduce, as is noted in Exercise 7.6. Finally, for the special case of first-order term instantiation, Exercise 7.7 guarantees that all variable renamings can be avoided (even without additional restrictions on the behavior of get_factor).

Example 6.3D showed that factorization can be greatly simplified by embedding an instantiation system in one that has a better set of constructs, while Exercise 7.8 and Example 7.3 investigate how this advantage carries over to unification algorithms. For many instantiation systems, cc_reduce can return multiequations that are clearly inappropriate (see Example 7.2B). This problem can be minimized by embedding the instantiation system in one with a richer type system, as shown by Proposition 7.10 and Example 7.2C.

Exercise 7.5. Assume that, after successful execution of restricts(σ, c, T), σ is always a weakening for c. Then, after successful execution of unify(δ, $\{v \approx t\}$), $\delta|$var(t) is a weakening for t.

Exercise 7.6.

A. If c is rigid and unifies_2c(σ, c, d) succeeds, we may assume $\sigma|$var(c) $= \epsilon$. In this case, $c\gamma$ is always a construct in cc_reduce.

B. The constructs of Example 6.5C are not rigid, but one can still give a construct-unification procedure satisfying the exit condition $c\sigma \in C$.

Exercise 7.7. Consider the special case of first-order term instantiation. Make the following assumptions:

i) the values of γ used by vv_reduce, vc_reduce, and cc_reduce are, respectively, v/u, c/v, and c/d.

ii) mulitequations are implemented as lists, with variables preceding nonvariables, and are processed from left to right, with union implemented as concatenation.

Then the unify loop satisfies the following invariants:

A. If size(E) > 1, then the last term of E occurs in a term of $\mathfrak{D}$.

B. If $t \in E \setminus V$ and t does not occur in a term of $\mathfrak{D}$, then t is the first term of E and t is a construct.

C. cdm(δ) $\subseteq$ var($\mathfrak{D}$) $\cup$ ($\cup\mathfrak{E}$) .

Hence, after unify(σ, $\mathfrak{D}$), cdm(σ) $\subseteq$ var($\mathfrak{D}$).

Exercise 7.8. Assume that $\mathfrak{B}$ is an ideal in $\mathfrak{C}$ (i. e., $t \in O^B$ implies $t\sigma \in O^B$); assume $\mathfrak{s}$ is a term system of $\mathfrak{B}$. If σ unifies $\mathfrak{s}$, then $\sigma|\mathrm{var}(\mathfrak{s}) \in S^B$. Thus, if unification$_1$ is supplied with support routines for $\mathfrak{C}$, and if unify is modified in such a way as to return $\delta|\mathrm{var}(\mathfrak{D})$, the resulting algorithm will return unifiers in S^B, whenever $\mathfrak{D}$ is a term system in $\mathfrak{B}$.

Exercise 7.9. Assume $\mathcal{A}$ is an equivalent subsystem of $\mathfrak{B}$ and $\mathfrak{C}$ is the extension of $\mathcal{A}$ induced by a type theory containing the intrinsic theory for $\mathcal{A}$; assume every primitive nonvariable belongs to $\mathcal{A}$. Let s, t be terms in $\mathfrak{B}$; let η be a slight weakening (in the sense of Exercise 5.6B).

A. If $\tau\eta$ unifies $s \approx t$, then so does τ. Consequently, the set of all $\mathfrak{B}$-unifiers of $s \approx t$ is complete in $\mathfrak{C}$.

B. Let Γ be a complete set of $\mathfrak{B}$-unifiers for $s \approx t$. A complete set of unifiers for $s\eta \approx t\eta$ is given by all substitutions of the form $(\eta^{\vee}\tau\rho)|\mathrm{var}(s\eta \approx t\eta)$, where $\tau \in \Gamma$. Moreover, we may assume $\mathrm{dom}(\rho) \subseteq \{v \mid \text{for some } u \in \mathrm{dom}(\eta),\ u\tau = v\}$.

C. If terms are weakly restrictable, then we may take ρ to be a weakening in Assertion B.

Proposition 7.10. Let $\mathfrak{B}$ be an effectively supported instantiation system with the following properties:

- The construct basis allows construct interpolation (see Exercise 6.6).
- Every construct c belongs to a smallest instance type T_c.
- factors(c, ϕ, t) succeeds, if t is a nonvariable, and on exit, c is a least-general construct for t.
- No variable is an instance of a nonvariable.

Then, $\mathfrak{B}$ may be fully embedded in an effectively supported instantiation system $\mathfrak{C}$ in such a way that, when using $\mathfrak{C}$, unify has the following properties, for any unifiable term system $\mathfrak{D}$ of $\mathfrak{B}$:

A. Every new multiequation output by a successful reduction step is either a fragment of its input multiequation, or is an instance of a unifiable *atomic* multiequation of the form $a \approx c$, where a is either a variable or a construct.

B. Any unifier returned by unify using support routines for $\mathfrak{C}$ belongs to $\mathfrak{B}$.

Proof. Define a *canonical* construct to be one returned by factors. Let C^B be the construct basis consisting of all canonical constructs. Let $C \subseteq C^B$ be a skeletal construct basis for $\mathfrak{B}$. Let $\mathfrak{B}'$ be an equivalent extension of $\mathfrak{B}$ containing infinitely many new variables of every type. Let θ be an inner isomorphism from $\mathfrak{B}'$ to $\mathfrak{B}$. Let $\mathfrak{C}$ be the extension of $\mathfrak{B}$ induced by the following type theory extending the intrinsic theory for $\mathfrak{B}$: The set of sorts contains one new sort I_c, for each construct c. The declared variables of the theory are given by the schema $u \in I_c$, for infinitely man new u such that $T_{u\theta} = T_c$. The typing axioms are given by, $c \in I_c$, for each I_c.

Now, C^B is already a construct basis for $\mathcal{C}$, because every primitive nonvariable of the type theory is a construct c in $\mathfrak{B}$. However, we need a larger basis, C^C, obtained by adding all terms of the form $c\eta$, where c is a $\mathfrak{B}$-construct and η is a slight weakening for c. We assume variables in I_c are of the form $\langle u, I_c \rangle$, where $T_{u\theta} = T_c$. Notice that one can sharpen Exercise 5.9B: every term is of the form $s = t\eta$, where $t = s(\theta|var(s) \setminus V^B)$, and $\eta = \theta^v|dom(\eta)$. Instantiation and related operations are computable in $\mathcal{C}$, because they are in $\mathfrak{B}$ and $\mathfrak{B}'$.

We next extend sub_both to a corresponding support procedure sub_both^C. There are essentially two new cases. In the first, sub_both^C(T, I_c, $T^B{}_v$) makes a pair of calls of the form restricts(ϕ, c, $T^B{}_v$); factors(e, α, $c\phi$), and returns I_e. This works, because, if $v \in V^B$ and $I_c \cap T^C{}_v \neq \emptyset$, then $c\sigma \in T^B{}_v$, for some σ; in this case, the construct interpolation property guarantees some d such that $v \leq d \leq c\sigma$, so that $d \leq e \leq c\sigma$, because e is least-general. In the other new case, sub_both^C(T, I_c, I_d) makes the calls unifies_2c(γ, c, d); factors(e, α, $c\gamma$), and returns I_e. The returned value of e is such that $I_e \subseteq I_c \cap I_d$, because e is a least-general construct for a common instance of c and d.

The procedures restricts^C and unifies_2c^C rely on the following *push_out* procedure, where τ, $\eta \in S^C$, η is variable-valued, $W \subseteq V^B$, and $\sigma \in S^B$:

```
procedure push_out(τ: out; W, η, σ: in) -- exit ητ ≥ σ [W]
is U := ∅;
begin
   for v in W loop
      if vσ ∈ V^C then
         sub_both^C(T, T_vη, T_vσ);
         pick any w ∈ V^C \ U so that T_w = T;
         τ(vη) := w; U := U ∪ {w};
      else if vη ≤ vσ then τ(vη) := vσ;
      else raise clash_failure end if;
   end loop
end push_out;
```

The fact that push_out computes a substitution follows from the above definition of sub_both^C. The motivation for raising clash_failure in the last case comes from the last assertion in Exercise 5.9B. If $v \in V^B$ and $c \in C^B$, the computation for restricts^C(τ, $c\eta$, $T^C{}_v$) consists of the pair of calls restricts(σ, c, $T^B{}_v$); push_out(τ, var(c), η, σ). Appropriateness follows from Exercise 5.9B. Similarly, the computation for restricts^C(τ, $c\eta$, I_d) consists of the pair of calls unifies_2c(σ, c, d); push_out(τ, var(c), η, σ). Finally, the computation for unifies_2c^C(τ, $c\eta$, $d\eta$) consists of the pair unifies_2c(σ, c, d); push_out(τ, var(c $\approx$ d), η, σ).

Before defining factors^C, we let factors′ be the corresponding support procedure for $\mathfrak{B}'$, and let factors″(d, β, s) be the computation

$$\text{factors}'(e, \gamma, s);\ \mu := \theta|\text{var}(e) \setminus V^B;\ d := e\mu;\ \beta := \mu^{\vee} \circ \gamma.$$

The purpose of factors″ is to find a $\mathfrak{B}$-construct for a nonvariable s of e, without having to first put s in the form s = tη. Finally, we have,

```
procedure factorsC(c, α: out; s: in)
is d; e; β; γ; η := ε; T; U := ∅;
begin
   factors"(d, β, s);
   for v in var(d) loop
      if vβ ∈ VC then
         pick any w ∈ V \ U with w equivalent to vβ; U := U ∪ {w};
      else factors(e, γ, vβ);
         pick any w ∈ V \ U with w of type Ie;  U := U ∪ {w};
      end if;
      η(v) := w
   end loop;
   c := dη; α := (η∨β)|var(c);
end factorsC;
```

The final substitution in factors^C is well-typed, as a result of Exercise 5.9B.

To prove Conclusion A, we must show that the multiequations produced by vc_reduce in $\gamma|\text{var}(c) \approx \alpha$ and by cc_reduce in $\gamma \approx \alpha + \beta$ are all instances of unifiable atomic multiequations. We consider only vc_reduce, as the other case is similar. We need to show that each $v\gamma \approx v\alpha$ has a unifiable atomic part; it suffices to consider the case where $v \in \text{dom}(\gamma) \cap \text{dom}(\alpha)$. From the definition of factors^C, we see that either $v\alpha/v$ is a renaming or $v\alpha$ is a nonvariable. In the latter case, $v \in I_c$, for some c. If $v \in I_c$, then $v\gamma \in I_c$. This implies that either $v\gamma/v$ is a renaming or $v\gamma$ is an instance of c. In the former case, $v\gamma \approx v\alpha$ itself is unifiable; in the latter case, $v\gamma \approx v\alpha$ has an atomic factorization of the form $c \approx c$.

Conclusion B follows from the following invariants:

$$\text{if } v \in E \setminus V^B \text{ and } E \in \mathcal{E}, \text{ then } E \setminus V^C \neq \emptyset.$$
$$\text{if } t \in \cup\mathcal{E}, \text{ and } v \in \text{var}(t) \setminus V^B, \text{ then } v \in \cup\mathcal{E}.$$
$$\text{cdm}(\delta) \subseteq V^B \cup (\cup\mathcal{E}).$$

Their validation is straightforward, but somewhat tedious, and is left as an exercise for the reader. □

EXAMPLES

We will use the following legend in presenting execution traces for the unify procedure:

- A : variables and constructs (*atoms*) to be unified in a reduction step.
- α : substitutions returned by get_factors.
- γ : unifiers computed during a reduction step.
- θ : portion of γ that gets composed with δ.
- σ : the final unifier.
- $\mathfrak{F}$: *fragment* systems of the form $\gamma|\text{var}(c) \approx \alpha$ or $\gamma \approx \alpha + \beta$.
- $\mathfrak{S}$: value of $\mathfrak{S}$ after a merge step.
- E : multiequation selected for reduction (and then successively reduced).

7.1. *Unification of First-Order Terms.* To see how unify handles first-order terms, consider a multiequation E of the form $\mathbf{x} \approx \mathbf{f}(r, s) \approx \mathbf{f}(r', s')$, and let $\mathfrak{D} = \{E\}$. The first few steps in the computation are,

A : $\mathbf{x} \approx \mathbf{f}(\mathbf{u}, \mathbf{v})$;	α : $r/\mathbf{u} + s/\mathbf{v}$;
γ : $\mathbf{f}(\mathbf{u}, \mathbf{v})/\mathbf{x}$;	θ : $\mathbf{f}(\mathbf{u}, \mathbf{v})/\mathbf{x}$;
A : $\mathbf{f}(\mathbf{u}, \mathbf{v}) \approx \mathbf{f}(\mathbf{u}', \mathbf{v}')$;	α : $r'/\mathbf{u}' + s'/\mathbf{v}'$;
γ : $\mathbf{u}/\mathbf{u}' + \mathbf{v}/\mathbf{v}'$;	$\mathfrak{F}$: $\{\mathbf{u} \approx r;\ \mathbf{v} \approx s;\ \mathbf{u} \approx r';\ \mathbf{v} \approx s'\}$.

$\mathfrak{S}$: $\{\mathbf{u} \approx r \approx r';\ \mathbf{v} \approx s \approx s'\}$.

. . . .

7.2. *Associative Unification.* Consider first-order term instantiation modulo associativity; let $\mathfrak{D} = \{\mathbf{x} + \mathbf{a} \approx \mathbf{a} + \mathbf{x}\}$. We suppress indication of equivalence classes for the sake of notational brevity.

A. *Direct Application.* A possible unify computation is,

1) A : $\mathbf{x} + \mathbf{v} \approx \mathbf{v}' + \mathbf{x}'$; $\quad\alpha$: $\mathbf{a}/\mathbf{v} + \mathbf{a}/\mathbf{v}' + \mathbf{x}/\mathbf{x}'$;
   γ : $(\mathbf{v}' + \mathbf{z})/\mathbf{x} + (\mathbf{z} + \mathbf{v})/\mathbf{x}'$; $\quad\theta$: ϵ;
   $\mathfrak{F}$: $\{\mathbf{v} \approx \mathbf{a};\ \mathbf{v}' \approx \mathbf{a};\ \mathbf{x} \approx (\mathbf{z} + \mathbf{v});\ \mathbf{x} \approx (\mathbf{v}' + \mathbf{z})\}$.

2) $\mathfrak{S}$: $\{\mathbf{v} \approx \mathbf{a};\ \mathbf{v}' \approx \mathbf{a};\ \mathbf{x} \approx (\mathbf{z} + \mathbf{v}) \approx (\mathbf{v}' + \mathbf{z})\}$;
   E : $\mathbf{x} \approx (\mathbf{z} + \mathbf{v}) \approx (\mathbf{v}' + \mathbf{z})$. — $\text{base}_{\mathfrak{S}} = \{E\}$
   A : $\mathbf{x} \approx (\mathbf{z} + \mathbf{v})$; $\quad\alpha$: ϵ;
   γ : $(\mathbf{z} + \mathbf{v})/\mathbf{x}$; $\quad\theta$: $(\mathbf{z} + \mathbf{v})/\mathbf{x}$;

3) A : $(\mathbf{z} + \mathbf{v}) \approx (\mathbf{v}' + \mathbf{z}')$; $\quad\alpha$: $\mathbf{z}/\mathbf{z}'$;
   γ : $\mathbf{z}/\mathbf{v}' + \mathbf{v}/\mathbf{z}'$; $\quad\theta$: ϵ;
   $\mathfrak{F}$: $\{\mathbf{z} \approx \mathbf{v};\ \mathbf{z} \approx \mathbf{v}'\}$.

4) $\mathcal{E}$: {z ≈ v ≈ v′ ≈ a ≈ a};
A : z ≈ v, v ≈ v′, v′≈ a, a ≈ a;
γ : v/z, v′/v, a/v′.
θ : v/z, v′/v, a/v′; $\mathcal{F}$: ∅.

8) $\mathcal{E}$: ∅.
σ : ((z + v)/x)(v/z)(v′/v)(a/v′).
σ|var($\mathfrak{D}$) : (a + a)/x. — final result.

Observe that Step (3) contains a multiequation similar to the original input. Consequently, this step could have duplicated the action of Step (1) by splitting the variable **z**. As a result, the possible unifiers for this example are those substitutions of the form **(a + ... + a)/x**. This same observation also shows that unify can run forever, in an apparent effort to substitute the infinite expression **(a + a + ...)** for **x**.

B. *Direct Application.* Step (1) of the above computation fortuitously splits **x** and **x′** instead of **v** and **v′**. For this instantiation system, the unify procedure has some excess nondeterminacy, as can be seen from the following unsuccessful computation:

1) A : x + v ≈ v′ + x′; α : a/v + a/v′ + x/x′;
γ : (z + x′)/v + (x + z)/v′; θ : ϵ;
$\mathcal{F}$: {x′ ≈ x; a ≈ (z + x′); a ≈ (x + z)}.

3) $\mathcal{E}$: $\mathcal{F}$.
E : (x ≈ x′); A : E; α : ϵ; γ : x/x′; θ : γ; $\mathcal{F}$: ∅.

4) $\mathcal{E}$: {a ≈ (z + x′); a ≈ (x + z)}.
E : (a ≈ (z + x′)). — $base_{\mathcal{E}} = \mathcal{E}$
A : E; α : ϵ; γ : clash_failure. — final result

The failure in Step 4 results from a poorly chosen atomic unifier in Step 1. This problem of producing poorly chosen subgoals was noticed by Stickel, in his commutative-associative unification procedure [STICK82]. He corrected the problem with an optimization whose net effect is similar to the one illustrated in Part C.

C. *After Addition of Construct-types.* We now consider the same unification problem in a larger system obtained by the extension described in the proof of Proposition 7.10. The new instance types include I_+ and I_a. The first factorization step proceeds as in Parts A and B, but now the declared type of the variable **v** is I_a. In this way, the extended factors procedure indicates that **v** must instantiate to a term that is not an instance of x + y. The computation in Step 1 of Part B cannot happen, because the function γ would not be substitution.

7.3. *Unification of First-order Terms and Formulas.* As in Example 6.3D, we regard (pure) first-order terms and formulas as an ideal subsystem of term-schemas and formula schemas. We will give a unification procedure for pure terms and formulas by giving suitable atomic factorizations of formulas in the larger system of all term- and formula-schemas.

A. *Factorization.* The interesting case is a multiequation involving quantified formulas, where all formulas have the same quantifier. For simplicity, we consider a two-formula multiequation. Using α-conversion we may put the multiequation in the form $[(\exists\ b)(\phi(b/\mathbf{z}))] \approx [(\exists\ b)(\psi(b/\mathbf{z}))]$, for some bound name b. A good factorization is,

A: $[(\exists\ b)(\boldsymbol{\alpha})] \approx [(\exists\ b)(\boldsymbol{\beta})]$; α: $[\phi(b/\mathbf{z})]/\boldsymbol{\alpha} + [\psi(b/\mathbf{z})]/\boldsymbol{\beta}$.

B. *Unification.* If factorization is performed as in Part A, there are no interesting cases. In particular, unifies_2c is applied only to multiequations of the form $[(\exists\ b)(\boldsymbol{\alpha})] \approx [(\exists\ b)(\boldsymbol{\beta})]$, in which case, $\boldsymbol{\alpha}/\boldsymbol{\beta}$ is a most general unifier.

C. *Un-optimized Factorization and Unification.* If we just perform factorization as in get_factor, without appropriate renamings of bound variables, then unifies_2c must deal with multiequations such as $[(\exists\ \mathbf{a})(\boldsymbol{\alpha})] \approx [(\exists\ \mathbf{b})(\boldsymbol{\beta})]$. The requisite complete set of unifiers for this multiequation consists of all substitutions of the form $[\psi(\mathbf{a}/\mathbf{z})]/\boldsymbol{\alpha} + [\psi(\mathbf{b}/\mathbf{z})]/\boldsymbol{\beta}$, where ψ varies over formula schemas having no metavariables other than $\mathbf{z}$. This example suggests that *common-construct* factorizations for multiequations may be generally preferable.

7.4. *Nontermination in First-order Systems.* Consider the subsystem of first-order term instantiation given by,

$\mathcal{T} = \{S, T_1, T_2, ...\}$; $S \supseteq T_1 \supseteq T_2$ ... ;
$\mathbf{x}, \mathbf{y}, \mathbf{x}', \mathbf{y}', \mathbf{F}(\mathbf{x}, \mathbf{y}) \in S$;
$\mathbf{z}_i, \mathbf{z}'_i, \mathbf{F}(\mathbf{z}_i, \mathbf{z}_i), \mathbf{F}(\mathbf{F}(\mathbf{x}, \mathbf{y}), \mathbf{F}(\mathbf{z}_{i+1}, \mathbf{z}_{i+1})) \in T_i$.

Notice that unification may be performed on $\mathbf{z}_i \approx \mathbf{F}(\mathbf{x}, \mathbf{y})$, giving the unifier $\mathbf{F}(\mathbf{x}', \mathbf{y}')/\mathbf{x} + \mathbf{F}(\mathbf{z}_{i+1}, \mathbf{z}_{i+1})/\mathbf{y} + \mathbf{F}(\mathbf{F}(\mathbf{x}', \mathbf{y}'), \mathbf{F}(\mathbf{z}_{i+1}, \mathbf{z}_{i+1}))/\mathbf{z}_i$. In the following example, the given input leads to an infinite computation:

E : $\mathbf{z}_2 \approx \mathbf{F}(\mathbf{x}, \mathbf{y}) \approx \mathbf{F}(\mathbf{z}_2, \mathbf{z}_2)$. — original input
A : $\mathbf{z}_2 \approx \mathbf{F}(\mathbf{x}, \mathbf{y})$; $\alpha : \epsilon$;
γ : $\mathbf{F}(\mathbf{F}(\mathbf{x}', \mathbf{y}'), \mathbf{F}(\mathbf{z}_3, \mathbf{z}_3))/\mathbf{z}_2 + \mathbf{F}(\mathbf{x}', \mathbf{y}')/\mathbf{x} + \mathbf{F}(\mathbf{z}_3, \mathbf{z}_3)/\mathbf{y}$;
θ : $\mathbf{F}(\mathbf{F}(\mathbf{x}', \mathbf{y}'), \mathbf{F}(\mathbf{z}_3, \mathbf{z}_3))/\mathbf{z}_2$.
$\mathcal{F}$: $\{\mathbf{x} \approx \mathbf{F}(\mathbf{x}', \mathbf{y}'); \mathbf{y} \approx \mathbf{F}(\mathbf{z}_3, \mathbf{z}_3)\}$;

E : $\mathbf{F}(\mathbf{F}(\mathbf{x}', \mathbf{y}'), \mathbf{F}(\mathbf{z}_3, \mathbf{z}_3)) \approx \mathbf{F}(\mathbf{z}_2, \mathbf{z}_2)$;
A : $\mathbf{F}(\mathbf{x}'', \mathbf{z}_3') \approx \mathbf{F}(\mathbf{z}_2, \mathbf{z}_2')$;
α : $\mathbf{F}(\mathbf{x}', \mathbf{y}')/\mathbf{x}'' + \mathbf{F}(\mathbf{z}_3, \mathbf{z}_3)/\mathbf{z}_3' + \mathbf{z}_2/\mathbf{z}_2'$;
γ : $\mathbf{z}_2/\mathbf{x}'' + \mathbf{z}_3'/\mathbf{z}_2'$;
$\mathcal{F}$: $\{\mathbf{z}_2 \approx \mathbf{F}(\mathbf{x}', \mathbf{y}'); \mathbf{z}_3' \approx \mathbf{F}(\mathbf{z}_3, \mathbf{z}_3); \mathbf{z}_3' \approx \mathbf{z}_2\}$;

$\mathcal{S}$: $\{z_2 \approx z_3' \approx F(x', y') \approx F(z_3, z_3);\ x \approx F(x', y');\ y \approx F(z_3, z_3)\}$.
θ : $F(x', y')/x,\ F(z_3, z_3)/y,\ z_3'/z_2;\ \ldots$
E : $z_3' \approx F(x', y') \approx F(z_3, z_3)$.

. . . . . .

Thus, for each i, unify works with z_i to create a similar problem involving z_{i+1}.

7.5. *Merging and Reduction.* Jaffar modified Martelli–Montanari reduction by performing *pure fragmentation*, without attempting to isolate a portion of the unifier [JAFF84]. This form of reduction is sound, even if not applied to base multiequations. The necessary modification for vc_reduce is given below, and the modification for vv_reduce is similar:

```
procedure vc_reduce(E, 𝒮, ℱ, U: in out)
is v; t; c; α, γ, ψ;
begin
    pick v ∈ E; pick t ∈ E \ V;
    get_factor(c, α, t, ∅, U);
    restricts(ψ, c, T_v);
    γ := cψ/v + ψ; rename_cdm(γ, var(v ≈ c), U);
    𝒮 := 𝒮 \ {E}; E := E \ {t} U {cγ}; 𝒮 := 𝒮 U {E} U (γ|var(c) ≈ α);
end vc_reduce;
```

Starting with first-order term instantiation, take the quotient system induced by the identities $C(S(x), y, z) = D(x, S(y), z)$ and $C(x, S(y), z) = D(x, y, S(z))$. Consider the term system

$\mathfrak{D}$: $\{[x] \approx [C(S(J), y_1, z_1)] \approx [C(x_2, S(y_2), L)];$
$[x] \approx [D(J, S(K), z_1')] \approx [D(J, y_2', S(L))]\ \}$.

A. We can see that $\mathfrak{D}$ is unifiable by merging, applying rewrite rules, guessing that $[x]\delta = [D(x_2, y_2, z_1)]$, and unify corresponding arguments:

After merging and applying rewrite rules:

$[x] \approx [D(J, S(y_1), z_1)] \approx [D(x_2, y_2, S(L))] \approx$
$[D(J, S(K), z_1')] \approx [D(J, y_2', S(L))]$.

After removal of $[D(x_2, y_2, z_1)]$ and merging of resulting fragments:

$\{\ x_2 \approx [J] \approx [J] \approx [J];\ y_2 \approx y_2' \approx [S(y_1)] \approx [S(K)];$
$z_1 \approx z_1' \approx [S(L)] \approx [S(L)]\ \}$.

At this point, it is obvious that the system is unifiable.

B. Reducing the first multiequation via pure fragmentation before merging can give,

$$A : ([x] \approx [C(u_1, y_1, z_1)] \approx [C(x_2, v_2, w_2)]);$$
$$\alpha : S(J)/u_1 + S(y_2)/v_2 + L/w_2;$$
$$\gamma : [C(u_1, y_1, z_1)]/x, u_1/x_2 + y_1/v_2 + z_1/w_2.$$
$$\mathcal{F} : \{u_1 \approx [S(J)]; u_1 \approx [S(y_2)]; z_1 \approx [L]; [x] \approx [C(u_1, y_1, z_1)]; u_1 \approx x_2\}.$$

Subsequent merging gives,

$$\mathcal{E} : \{u_1 \approx x_2 \approx [S(J)]; y_1 \approx [S(y_2)]; \; z_1 \approx [L]; [x] \approx [C(u_1, y_1, z_1)] \approx [D(J, S(K), z_1')] \approx [D(J, y_2', S(L))] \}.$$

This term system is not unifiable, and the execution must eventually fail: To unify the last multiequation, it is necessary to express some instance of $C(u_1, y_1, z_1)$ in terms of the function D. Only two possibilities are allowed by the term-equivalence relation. The first leaves z_1 unchanged, so that z_1 must eventually unify with both $[L]$ and $[S(L)]$. The second leaves u_1 unchanged, so that u_1 must unify with both $[J]$ and $[S(J)]$.

7.6. *Retyping and the Selection of Base Multiequations.* We perform merging before reduction, but continue to use fragmentation-reduction and defer recognition of base multiequations, as in Jaffar's algorithm.

A. Consider the order-sorted system given by,

$$\mathcal{T} = \{S_1, S_2, ..., T_1, T_2, ...\}; \; S_{i+1} \subseteq S_i \text{ and } T_{i+1} \subseteq T_i;$$
$$x_i, x'_i \in S_i; \; y_i, y'_i \in T_i;$$
$$F(x_i) \in T_j, \text{ if } j < i; \; G(y_i) \in S_j, \text{ if } j < i.$$

Constructs are those terms of the form $F(u)$, $G(v)$. Atomic unification achieves variable retyping via the typing axioms in the obvious way. Consider the nonunifiable term system $\mathcal{E} = \{x_1 \approx G(y_1); y_1 \approx F(x_1)\}$. If no effort is made to select base multiequations, alternate merging and reduction could fail to terminate as follows:

$$x_1 \approx G(y_1); \; y_1 \approx F(x_1) \qquad - \gamma = y_2/y_1 + G(y_2)/x_1$$
$$x_1 \approx G(y_2); \; y_2 \approx y_1 \approx F(x_1) \qquad - \gamma = x_3/x_1 + F(x_3)/y_1 + F(x_3)/y_2$$
$$x_3 \approx x_1 \approx G(y_2); \; y_2 \approx y_1 \approx F(x_3)$$

. . . . . .

Thus, for each $i \geq 1$, reduction works on y_i or x_i to create a similar problem for x_{i+1} or y_{i+1}, respectively.

B. Even if the term system is unifiable, a poor choice of the order in which multiequations are handled can lead to an exponential-time retyping activity. Consider the order-sorted system given by

$\mathcal{T} = \{T_M \mid M \subseteq \omega\}$; $T_M \subseteq T_N$ iff $M \supseteq N$.
For each $M \subseteq \omega$ and $i \in \omega$, variables, x_{ijM}, $y_{ijM} \in T_M$.
For each $k \in \omega$, a k-ary construct $F_k(x_{1\emptyset}, \ldots, x_{k\emptyset})$ that satisfies the typing schema $F_k(x_{1M_1}, \ldots, x_{kM_k}) \in T_N$, provided $N \subseteq M_1 \cap \ldots \cap M_k$.

Consider a term system of the form,

$$E_1 : x_{11\{1\}} \approx x_{12\emptyset} \approx x_{13\emptyset} \approx \ldots \approx x_{1n\emptyset} \approx F_1(x_{11\emptyset}).$$
$$E_2 : x_{22\{2\}} \approx x_{23\emptyset} \approx \ldots \approx x_{2n\emptyset} \approx F_2(x_{12\emptyset}, x_{22\emptyset}).$$
$$\cdot \quad \cdot \quad \cdot \quad \cdot \quad \cdot \quad \cdot$$
$$E_n : x_{nn\{n\}} \approx F_n(x_{1n\emptyset}, x_{2n\emptyset}, \ldots, x_{nn\emptyset}).$$

Each multiequation suggests a substitution that maps its variables to a weakening of its nonvariable. Retyping via atomic reduction may work on the E_j's by always choosing the first E_j that needs retyping. Subsequent merging of new variables will then cause all preceding E_i's to need retyping. The first three steps go as follows:

$$E_1 : x_{11\{1\}} \approx x_{12\emptyset} \approx x_{13\emptyset} \approx \ldots \approx x_{1n\emptyset} \approx F_1(y_{11\{1\}}); \quad y_{11\{1\}} \approx x_{11\emptyset}.$$
$$E_2 : x_{22\{2\}} \approx x_{23\emptyset} \approx \ldots \approx x_{2n\emptyset} \approx F_2(x_{12\{2\}}, y_{22\{2\}}); \quad x_{12\emptyset} \approx x_{12\{2\}};\ x_{22\emptyset} \approx y_{22\{2\}}.$$
$$E_1 : x_{11\{1\}} \approx x_{12\emptyset} \approx x_{12\{2\}} \approx x_{13\emptyset} \approx \ldots \approx x_{1n\emptyset} \approx F_1(x_{11\{1,2\}}); \quad x_{11\{1\}} \approx x_{11\emptyset}.$$

By way of contrast, if base multiequations are handled first, beginning with E_n, then each multiequation is handled only once, because the typing constraints imposed by a given multiequation can only propagate to multiequations that come later in a triangular enumeration of the term system.

SECTION 8
TERM-IMPLEMENTATION AND COMPLETENESS

The algorithm given in Section 7 need not be implementable for some instantiation systems, because the var operator need not be computable (see Example 8.1). Even if it is, the algorithm need not be complete. If terms are not well founded, successive factorizations may fail to decompose a term in finitely many steps. Both of these problems can be addressed by applying a variant of the algorithm to *concrete* terms in the domain of a suitable homomorphism.

We consider a homomorphism g similar to the one discussed in Theorem 6.4. The constraints imposed on concrete terms are sufficiently weak as to allow direct support for associative instantiation and for α-instantiation via deBruijn indexes [BRUI72]. We begin by observing that g induces an *implementation* of basic unification concepts defined in Section 7.

A variant of the algorithm in Section 7 is then shown to be both *sound* and *complete relative to* g. This analysis also shows the extent to which completeness depends on nondeterminacy in the algorithm. In preparation for Section 9, we give a step-by-step analysis showing that the new algorithm is a *partial implementation* of the one from Section 7. Finally, we compare completeness relative to g to a similar but more simplistic notion of *tree-completeness*.

CONCRETE TERM IMPLEMENTATION

A technique for implementing data abstractions via first-order homomorphisms has been given by Guttag, Horrowitz, and Musser [GUTG76]. This technique can be extended to (non first-order) homomorphisms for instantiation systems. To perform the term-implementation, we first extend the function $g^{\sim}$ introduced in Section 5 to act on term systems and other *concrete entities*. We then define *images* of *abstractable* entities and *preimages* of *implementable* entities. The preimage concept provides a uniform approach to a variety of standard concepts including E-unification and the relation $\sigma \leq_E \tau$ [U]. Notice that terms such as "concrete" and "implementable" make sense only with respect to $g^{\sim}$. They will have appropriately different meanings for the implementation steps in Section 9.

A complicating factor in unifying concrete terms is that $base_{\mathfrak{s}}$ may be empty even though $g^{\sim}(\mathfrak{s})$ is unifiable. In this case, a new support procedure *simplify*(t) can be used to obtain an equivalent $\mathfrak{s}'$ such that $g^{\sim}(\mathfrak{s}') = g^{\sim}(\mathfrak{s})$ and $base_{\mathfrak{s}'} \neq \emptyset$. The purpose of simplify(t) is to eliminate *extraneous* variables belonging to $g^{\sim}(var(t)) \setminus var(g(t))$. Its formulation relies implicitly on the assumption that every variable has a variable-free instance. As indicated by Proposition 5.10, this assumption is quite mild.

Table 8.1 introduces the package *unification*$_2$ as a variant of unification$_1$. It uses the new simplify support procedure and recasts the exit conditions for unifies_2c and unify so that they now involve unification relative to g. Internally, there is only one change to the algorithm: instead of raising clash_failure when $base_{\mathcal{S}} = \emptyset$, unify now calls *break_cycle*, a new procedure that replaces $\mathcal{S}$ with an equivalent concrete term system and then raises cycle_failure, if the new term system fails to contain base multiequations. The resulting *unify*$_2$ algorithm is sound for essentially the same reasons as the original.

Throughout the rest of this paper, we assume that $\mathfrak{B}$ has weakly-restrictable, acyclic terms, that $\mathcal{A}$ has well-founded terms and variable-preserving substitutions, that every term in $\mathcal{A}$ has a variable-free instance, and that $g : \mathcal{A} \to \mathfrak{B}$ is a variable-preserving, type-preserving quotient homomorphism. Thus, $\mathcal{A}$ also has weakly-restrictable terms. We also assume g maps a fixed construct basis for $\mathcal{A}$ to a corresponding construct basis for $\mathfrak{B}$. Thus, g maps nonvariables to nonvariables.

The full definition of $g^{\sim}$ is given as follows:

terms:	$g^{\sim}(t) = g(t)$.
substitutions:	$g(v)g^{\sim}(\sigma) = g(v\sigma)$.
sets of terms:	$g^{\sim}(X) = \{g(t) \mid t \in X\}$.
integers, boolean values:	$g^{\sim}(x) = x$.
multiequations:	$g^{\sim}(E) = \{g(t) \mid t \in E\}$.
term systems:	$g^{\sim}(\mathcal{S}) = \{g^{\sim}(E) \mid E \in \mathcal{S}\}$.
(e. g., parameter) lists:	$g^{\sim}(x)_i = g^{\sim}(x_i)$.
program states:	$g^{\sim}(S).v = g^{\sim}(S.v)$, for any state-component v.
I/O relations:	$g^{\sim}(R) = \{\langle g^{\sim}(x), g^{\sim}(y)\rangle \mid \langle x, y\rangle \in R\}$.
functions:	$dom(g^{\sim}(f)) = g^{\sim}(dom(f))$ and
	$g^{\sim}(f)(g^{\sim}(x)) = g^{\sim}(f(x))$, provided f is abstractable.

A function f is *abstractable (on concrete values)* iff $g^{\sim}(x) = g^{\sim}(x')$ implies $f(x) = f(x')$, for all x and x′ in the domain of both f and g. We extend $g^{\sim}$ to functions with multiple arguments by treating such functions as having single arguments that happen to be lists.

The well-definedness of $g^{\sim}$ is not automatically guaranteed. One may justify its well-definedness by using a language semantics in which no entity can be more than one of the following: an integer, a boolean, a term, a function, a (multi)set [cf GUTM85, Section 1]. (Alternatively, one may regard $g^{\sim}$ as an overloaded operator whose value depends on both the explicitly passed argument and its type; in this case, one needs an explicit type-inference mechanism and a convention for handling arguments of unknown type.)

A *concrete entity* (respectively, *abstract entity*) is an element of the domain (range) of $g^{\sim}$. If x is a concrete entity, we refer to $g^{\sim}(x)$ as the *image* of x, and refer to x as an *implementation* of $g^{\sim}(x)$. We say f is *uniquely implementable (on abstract entities)* iff whenever $g^{\sim}(x) \in \mathrm{dom}(f)$, there is a unique y such that $g^{\sim}(y) = f(g^{\sim}(x))$. In this case, we define f_g, the *preimage* of f, as follows: $\mathrm{dom}(f_g) = \{x \mid g^{\sim}(x) \in \mathrm{dom}(f)\}$; $g^{\sim}(f_g(x)) = f(g^{\sim}(x))$, whenever $g^{\sim}(x) \in \mathrm{dom}(f)$.

If the range of f consists of integers, booleans, or return–codes, then f is uniquely implementable. If f is uniquely implementable, then f_g is abstractable and $g^{\sim}(f_g)$ is just f restricted to abstract values. If E is the congruence relation on first-order terms induced by a quotient homomorphism g, then $s =_g t$ iff $s\ E\ t$; σ unifies$_g$ $\mathcal{S}$ iff σ E–unifies $\mathcal{S}$; and $\sigma \leq_g \tau\ [U]$ iff $\sigma \leq_E \tau\ [U]$.†

We present implementations by explicitly presenting each syntactic alteration to the algorithm being implemented, using the form *pattern ==> replacement*. The *scope* where each replacement is performed is indicated by embedding that replacement in the partially elided specification to be implemented.

Let *unify*$_1$ = unification$_1$.unify. Let *unification*$_2$ be the package described in Table 8.1; let *unify*$_2$ = unification$_2$.unify.

Theorem 8.1 (soundness$_g$). The unify$_2$ procedure is sound$_g$, given the above–announced constraints on g. Moreover, the assertions of Table 7.1 carry over to unify$_2$, with the following changes:

unify exit: δ unifies$_g$ $\mathcal{D}$.
cc_reduce exit: ... γ unifies$_g$ $c \approx d$, ...
break_cycle exit: $\mathcal{S}' =_g \mathcal{S}$ and $\mathrm{var}(\mathcal{S}') \subseteq \mathrm{var}(\mathcal{S})$.

Proof. The exit conditions in Table 7.1 that do not relate to unification also hold for the procedures of unification$_2$, because their proofs did not depend on the exit condition for unifies_2c. The exit condition for break_cycle follows directly from its definition. The exit conditions and loop invariants relating to unification are proved just as in Theorem 7.4, replacing unification with unification$_g$ and, occasionally, (=) with ($=_g$). □

COMPLETENESS

With suitable constraints on $\mathcal{A}$, any unifier$_g$ of a unifiable$_g$ term system is both less-general$_g$ and *heavier* than one that can be returned by unify$_2$. Moreover, we can identify the specific nondeterministic steps of the unify$_2$ algorithm that must be controlled in order to achieve completeness. The completeness concept that we will verify thus distinguishes between two kinds of nondeterminacy. *Control* nondeterminacy ensures completeness of the algorithm, whereas *design*

† The establishment of a uniform notation has, unfortunately, led to certain departures from tradition.

Pattern		Replacement
Generic package *$unification_1$*	==>	**Generic package** *$unification_2$*
...		
procedure *unifies_2c* ...		
... σ unifies c ≈ d ...	==>	... σ $unifies_g$ c ≈ d ...
...		
	==>	**procedure** *simplify*(t: in out);
		-- exit $t' =_g t$ and $var(t') \subseteq var(t)$;
is		
procedure *unify* ...		
-- exit σ unifies $\mathfrak{D}$;	==>	-- exit σ $unifies_g$ $\mathfrak{D}$;
...		
Package body *$unification_1$*	==>	**Package body** *$unification_2$*
procedure *unify* ...		
is ...		
raise cycle_failure;	==>	break_cycle($\mathfrak{S}$);
...		
end unify;		
...		
	==>	**procedure** *break_cycle*($\mathfrak{S}$);
		is s; E; $\mathfrak{F}$: constant := $\mathfrak{S}$;
		begin
		for F in $\mathfrak{F}$ loop
		E := F;
		for t in F \ V loop
		s := t; simplify(s);
		E := E \ {t} ∪ {s};
		end loop;
		$\mathfrak{S}$:= $\mathfrak{S}$ \ {F} ∪ {E};
		end loop;
		if $base_{\mathfrak{S}}$ = ∅ then raise cycle_failure;
		end if;
		end break_cycle;
end $unification_1$;	==>	end $unification_2$;

Table 8.1: Term Implementation

nondeterminacy (also called "don't care" nondeterminacy) is used to suppress implementation details. In the algorithm, control nondeterminacy is encapsulated in the support procedures, sub_both, restricts, unifies_2c, and simplify, whereas design nondeterminacy is found in the factors support procedure and in the "pick any" and "select any" statements. Control nondeterminacy is discussed further at the end of this subsection.

We formalize control nondeterminacy in terms of *feasibility* assertions that allow us to specify multiple, possibly incompatible goals. Feasibility assertions are superficially similar to entry/exit assertions, with several major differences. In addition to I/O parameters, feasibility conditions may contain other free variables representing control information needed to select and achieve particular goals. Successful termination is implicitly asserted as part of the conclusion. Output parameters are treated as being existentially quantified.

For loop statements, proving termination involves showing that some well-founded *measure* can decrease with each pass of the loop. Soundness of this and other necessary proof techniques is not explicitly considered.

Let $wt(t)$ be an arbitrary weight function for $\mathcal{A}$. Define $s =_g \leq_{wt} t$ iff $s =_g t$ and $wt(s) \leq wt(t)$. Define $\sigma \equiv_g \tau$ [U] iff $v\sigma =_g v\tau$, for all $v \in U$; define $\sigma \equiv_g \leq_{wt} \tau$ [U] iff $v\sigma =_g \leq_{wt} v\tau$, for all $v \in U$. Define $\sigma \leq_g \leq_{wt} \tau$ [U] iff $\sigma\rho \equiv_g \leq_{wt} \tau$ [U], for some ρ.

For any concrete multiset E of terms, *meas*(E) is the multiset of all wt(t), with $t \in E$. Measures of multisets are ordered lexicographically, in the sense that meas(E) < meas(F) iff there is an $n \in$ meas(F) such that n has strictly fewer occurrences in meas(E) than in meas(F), and every ordinal greater than n has equally many occurrences in both meas(E) and meas(F). Finally, we define *meas*($\mathcal{E}$) to be the multiset, {meas(E) | $E \in \mathcal{E}$}. Measures of term systems are also ordered lexicographically.

Exercise 8.2. Assume wt is a weight function for $\mathcal{A}$.

A. Measures of multiequations and term systems are well-founded.
B. meas(merge($\mathcal{F}$)τ) $\leq$ meas($\mathcal{F}\tau$).
C. If wt is additive and $\sigma \equiv_g \leq_{wt} \tau$ [var($\mathcal{E}$)], then meas($\mathcal{E}\sigma$) $\leq$ meas($\mathcal{E}\tau$).
D. If meas(F) < meas({t}), then meas(E $\cup$ F) < meas(E $\cup$ {t}).
E. If meas($\mathcal{F}$) < meas({F}), then meas($\mathcal{E}$ $\cup$ $\mathcal{F}$) < meas($\mathcal{E}$ $\cup$ {F}).

For a given I/O relation R(x, y), we write

R(x, y): *feasibly, if* p(x, z), *then* q(x, z, y)

to mean that, if p(x, z), then, for some y, R(x, y), q(x, z, y), and y is successful. For deterministic procedures and single-valued I/O relations, feasibility coincides with total correctness.

To say that a unification algorithm with output σ and input $\mathcal{D}$ is *complete*$_g$ is to say that, feasibly, if τ unifies$_g$ $\mathcal{D}$, then σ is more-general$_g$ than τ on var($\mathcal{D}$). In Theorem 8.3, the feasibility condition for unifies_2c cannot be weakened to completeness$_g$ (see Example 8.2).

Theorem 8.3 (completeness$_g$). Assume wt is an additive weight function for $\mathcal{A}$. Add the following feasibility conditions to the specification of the support procedures for unification$_2$:

sub_both(T, T_1, T_2): feasibly, if $t \in T_1 \cap T_2$, then $t \in T$.
restricts(σ, c, T): feasibly, if $c\tau \in T$, then $\sigma \leq_g \leq_{wt} \tau$ [var(c)].
unifies_2c(σ, c, d): feasibly, if τ unifies$_g$ $c \approx d$,
then $\sigma \leq_g \leq_{wt} \tau$ [var($c \approx d$)].
factors(c, σ, t): $t \notin V$ implies successful termination.
simplify(t): feasibly, var(t′) = var$_g$(t) and wt(t′) $\leq$ wt(t).

Then unify$_2$(σ, $\mathfrak{D}$) is such that, feasibly, if τ unifies$_g$ $\mathfrak{D}$, then $\sigma \leq_g \leq_{wt} \tau$ [var($\mathfrak{D}$)]. In particular, unify$_2$ is complete$_g$.

Proof. Assume τ unifies $\mathfrak{D}$. Let U_0 be the value of U which results from execution of the statement pick any U with U $\supseteq$ var($\mathfrak{D}$). Consider the following loop invariant:

exists ρ such that ρ unifies$_g$ $\mathfrak{S}$ and $\delta\rho \leq_g \leq_{wt} \tau$ $[U_0]$.

If the loop terminates, letting $\rho = \epsilon$ shows $\delta \leq_g \leq_{wt} \tau$ $[U_0]$. This loop invariant holds trivially on entry to the loop, with $\rho = \tau$. We need to prove that, feasibly, the invariant is preserved on each pass, and, feasibly, the loop terminates. In giving the proof, we split the body of the unify$_2$ loop into two portions consisting of the first select statement and the ensuing reduction. We will at first assume the following two feasibility conditions concerning merging and reduction:

merge & break_cycle: feasibly, base$_{\mathfrak{S}'}$ $\neq \emptyset$.

reduction: feasibly, if ρ unifies$_g$ $\mathfrak{S}$, then there exists ρ', θ such that
$\delta' = \delta\theta$, $\theta\rho' \equiv_g \leq_{wt} \rho$ [U],
ρ' unifies$_g$ $\mathfrak{S}'$, and meas($\mathfrak{S}'\rho'$) < meas($\mathfrak{S}\rho$).

The first condition implies that selecting a next base multiequation is always feasible. The second condition and Exercise 8.2B imply that, feasibly, the measure of $\mathfrak{S}\rho$ always decreases, so that the loop eventually terminates, and we now show that the ρ' it provides can be used to propagate the loop invariant. It implies $\delta'\rho' = \delta\theta\rho' \equiv_g \leq_{wt} \delta\rho$ [U], by Exercise 6.7D, and $\delta\rho \leq_g \leq_{wt} \tau$ $[U_0]$, by loop induction. Since $U_0 \subseteq U$, these two facts show that the loop invariant can propagate. Hence, it remains to prove the above two feasibility conditions.

To prove the first, we consider the case where merging is selected. After merging, base$_{g^{\sim}(\mathfrak{S})}$ $\neq \emptyset$, by Proposition 7.2, since terms in $\mathfrak{B}$ are acyclic and $g^{\sim}(\rho)$ unifies $g^{\sim}(\mathfrak{S})$ by loop induction. If base$_{\mathfrak{S}}$ is still empty, then break_cycle is called, and it suffices to prove the following condition for break_cycle:

feasibly, if $\mathfrak{S}$ is merged and base$_{g^{\sim}(\mathfrak{S})}$ $\neq \emptyset$, then base$_{\mathfrak{S}'}$ $\neq \emptyset$.

It is easy to see, from the conditions on simplify, that $g^{\sim}(\mathcal{S}) = g^{\sim}(\mathcal{S}')$ and, feasibly, $var(t) = var_g(t)$, for all $t \in \bigcup\mathcal{S}$. Choose $E \in \mathcal{S}'$ so that $g^{\sim}(E) \in base_{g^{\sim}(\mathcal{S})}$. To see that $E \in base_{\mathcal{S}'}$, pick $F \in \mathcal{S}'$; we need to show $E \cap var(F \setminus V) = \emptyset$. If $v \in E$, $t \in F \setminus V$ and $v \in var(t)$, then $v \in var_g(t)$ by assumption. But then $g^{\sim}(v) \in g^{\sim}(E) \cap var(g(t))$, proving that $g^{\sim}(E) \notin base_{g^{\sim}(\mathcal{S})}$, a contradiction.

To prove the feasibility condition for reduction, assume ρ unifies$_g$ $\mathcal{S}$. The algorithm picks $E \in base_{\mathcal{S}}$, and thus ρ unifies$_g$ E. Assume, by restricting ρ, if necessary, that $dom(\rho) \subseteq U$. If vv_reduce, is selected, it picks $u \in E$ and $v \in E \setminus \{u\}$. By assumption, $u\rho = v\rho \in T_u \cap T_v$. Feasibly, sub_both chooses w so that $v\rho \in T_w$, and the appropriate value of θ is $w/u + w/v$.
Let $\rho' = \rho|(U \setminus \{u, v\}) + v\rho/w$. It is easy to see that $\theta\rho' = \rho\ [U]$, so that $\theta\rho' \equiv_g\leq_{wt} \rho\ [U]$ and ρ' unifies$_g$ $(\mathcal{S} \setminus \{E\}) \cup (E \setminus \{u, v\})$. Moreover, if $t \in E' \cap E$, then $t\rho' = t\rho =_g v\rho = w\rho'$, using $var(t) \cap E = \emptyset$, ρ unifies$_g$ $\mathcal{S}$, and choice of ρ'. Hence ρ' unifies$_g$ E'; hence ρ' unifies$_g$ $\mathcal{S}'$. Finally, $meas(\mathcal{S}'\rho') < meas(\mathcal{S}\rho)$, since $\mathcal{S}\rho$ is obtainable from $\mathcal{S}'\rho'$ by adding one more copy of $v\rho$ (= $w\rho'$) to E', by Exercises 8.2D and 8.2E.

If cc_reduce is selected, the algorithm picks s, t, c, d, α, and β; in this case, $\theta = \epsilon$. Notice that $(\alpha + \beta)\rho$ unifies$_g$ $c \approx d$, since ρ unifies$_g$ $s \approx t$. Let U_2 be the value of U after s and t have been factored. Feasibly, unifies_2c chooses γ so that $\gamma \leq_g\leq_{wt} (\alpha + \beta)\rho\ [var(c \approx d)]$; this remains true after renaming of $cdm(\gamma)$. Choose ξ so that $\gamma\xi \equiv_g\leq_{wt} (\alpha + \beta)\rho\ [var(c \approx d)]$, and $dom(\xi) \subseteq var(c \approx d) \cup cdm(\gamma)$. By construction, $cdm(\gamma) \cap U_2 \subseteq var(c \approx d)$. This and choice of ξ imply $dom(\xi) \cap (U_2 \setminus var(c \approx d)) = \emptyset$.

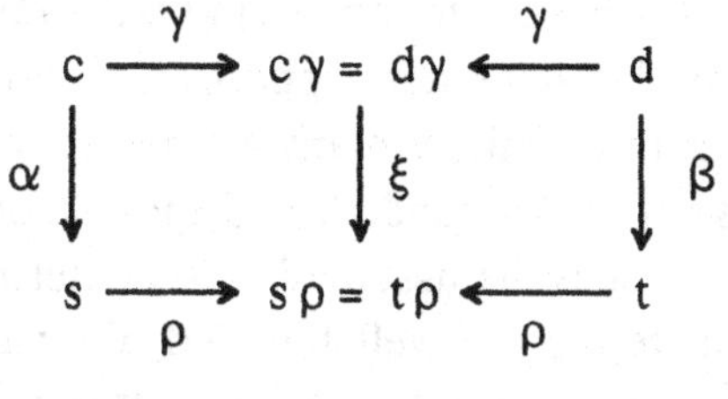

Let $\rho' = (\gamma\xi)|var(c \approx d) + \rho|(U \setminus var(c \approx d))$.

1. $\rho' \equiv_g\leq_{wt} \rho\ [U]$, since if $v \in U \cap var(c \approx d)$, then $v\rho' = v\gamma\xi =_g\leq_{wt} v(\alpha + \beta)\rho = v\rho$.
2. $\gamma\rho' \equiv \rho'\ [U_2]$, since $cdm(\gamma) \cap U \subseteq var(c \approx d)$, and $cdm(\gamma) \cap dom(\gamma) = \emptyset$.
3. $(\alpha + \beta)\rho \equiv_g (\alpha + \beta)\rho'\ [U_2]$, since $cdm(\alpha + \beta) \subseteq U$ and $\rho \equiv_g \rho'\ [U]$.
4. $\gamma\rho' \equiv_g (\alpha + \beta)\rho'\ [var(c \approx d)]$, by Step 2, choice of ρ', ξ, and Step 3.
5. If $r \in E \cap E'$, then $r\rho' =_g r\rho =_g t\rho = d(\alpha + \beta)\rho =_g d\gamma\xi = d\rho' =_g d\gamma\rho'$, using Step 1, ρ unifies E, choice of β, choice of ξ, choice of ρ', Step 2.
6. ρ' unifies $\mathcal{S}'$, by Steps 4 and 5.

7. $wt(d\gamma\rho') = wt(d\gamma\xi) \leq wt(d(\alpha + \beta)\rho) = wt(t\rho)$,
using Step 2 and choice of ρ', choice of ξ and Exercise 6.7D, choice of β.
8. $meas(\mathcal{E}'\rho') < meas(\mathcal{E}\rho)$, using Exercises 8.2C, 8.2D, and 8.2E.

Finally, if vc_reduce is selected, the algorithm picks v, t, c, and α so that $c\alpha = t$. Let $U_1 = U \cup var(c)$. We know $\alpha\rho$ unifies$_g$ $v \approx c$, since $v \notin dom(\alpha)$ and ρ unifies$_g$ $v \approx t$. Feasibly, restricts chooses ψ so that $\psi \leq_g \leq_{wt} \alpha\rho$ [var(c)]. Moreover, γ is constructed from ψ in such a way that $\gamma \leq_g \leq_{wt} \alpha\rho$ [var(c)], so we may choose ξ so that $\gamma\xi \equiv_g \leq_{wt} \alpha\rho$ [var(c)] and $dom(\xi) \subseteq var(c) \cup cdm(\gamma)$. By construction, $cdm(\gamma) \cap U_1 \subseteq var(c)$. This and choice of ξ imply $dom(\xi) \cap (U_1 \setminus var(c)) = \emptyset$.
Let $\rho' = \gamma\xi|var(c) + \rho|(U \setminus var(c \approx v))$; let $\theta = c\gamma/v$.

1. $\theta\rho' \equiv_g \leq_{wt} \rho$ [U]; in particular, $v\theta\rho' = c\gamma\gamma\xi =_g \leq_{wt} c\alpha\rho = t\rho$.
2. $\gamma\rho' \equiv \rho'$ $[U_1 \setminus \{v\}]$.
3. $\alpha\rho \equiv_g \alpha\rho'$ $[U_1 \setminus \{v\}]$, since $dom(\alpha) \subseteq U_1 \setminus \{v\}$ and $cdm(\alpha) \subseteq U \setminus \{v\}$.
4. $\gamma\rho' \equiv_g \alpha\rho'$ [var(c)].
5. If $s \in E \cap E'$, then $s\rho' =_g s\rho =_g t\rho = c\alpha\rho =_g c\alpha\rho' =_g c\gamma\rho'$.
6. $\theta\rho'$ unifies $\mathcal{E}'$, by Steps 4 and 5.
7. $wt(v\gamma\rho') = wt(c\gamma\rho') = wt(c\gamma\xi) \leq wt(c\alpha\rho) = wt(t\rho)$.
8. $meas(\mathcal{E}'\rho') < meas(\mathcal{E}\rho)$. □

Exercise 8.4

A. Assume terms in $\mathfrak{B}$ are well-founded and, in $\mathcal{A}$, wt = rank$_g$. Then wt is a weight function, $s =_g \leq_{wt} t$ iff $s =_g t$, and $\sigma \equiv_g \leq_{wt} \tau$ [U] iff $\sigma \equiv_g \tau$ [U].
B. Assume terms in $\mathfrak{B}$ are well-founded, and wt = rank$_g$. Then unify$_2$ is complete$_g$, even if $\mathcal{A}$ does not support an additive weight function.
C. Assume, in $\mathfrak{B}$, that only finitely many terms occur in any given term. Then terms are well-founded, and the var operator is computable. In this case, simplify may be assumed to be deterministic.
D. Assume, in $\mathfrak{B}$, that terms are well-founded, the var operator is computable, and substitutions are variable-preserving. In this case, we may take g to be the identity homomorphism and take simplify to be a no-op, so that unify$_1$ coincides with unify$_2$ and is, therefore, complete.

Exercise 8.5 Assume that, in $\mathcal{A}$, that substitutions are variable-preserving and each term has only finitely many subterms, so that unify$_1$, is complete.

A. If $\mathcal{A}$ has finite type intersections, then sub_both can be deterministic.
B. Assume constructs are *uniformly restrictable,* in the sense that for given c and v, there exists variable-valued η such that whenever $v \leq c\tau$, we have $v \leq c\eta$ and $\eta \leq \tau$ [var(c)]. If c and v have a common instance, then η is essentially unique. Hence, nondeterminacy is unnecessary in restricts.

C. Assume terms have unique quotients. If c is a common construct for $t_1, \ldots, t_n$, then the term systems $\{t_1 \approx \ldots \approx t_n\}$ and $(t_1/c \approx \ldots \approx t_n/c)$ have the same unifiers.

D. Assume, in addition, that terms are *decomposable* in the sense that any pair of unifiable terms has a common construct. Then, the observation in part C may be used to avoid unifies_2c in the definition of cc_reduce, eliminating this source of nondeterminacy as well.†

It should be clear from the proof of Theorem 8.3 that we have identified the minimum set of control points needed to prove completeness of the unify procedure. We have not, however, minimized the amount of control nondeterminacy. Indeed, the problems identified in Examples 7.2B and 7.3C both involve excess control nondeterminacy. We now briefly discuss yet another such problem.

It can easily happen in cc_reduce that $\mathrm{dom}(\gamma) \cap \mathrm{var}(\mathfrak{s}) \neq \emptyset$. If $v \in \mathrm{var}(\mathfrak{s})$ and $v\gamma$ is not equivalent to v, then $\gamma|\{v\}$ is a guessed fragment of the unifier to be returned that has been chosen by unifies_2c. During the course of a computation, unifies_2c can easily return several incompatible guesses. Thus, failure to treat $\gamma|\mathrm{var}(\mathfrak{s})$ as part of the emerging partial unifier leads to excess control nondeterminacy. This problem also exists in Algorithm 2 of Martelli et al., but is obscured by their liberal use of variable renamings.

This problem can be addressed by accumulating $\gamma|\mathrm{var}(\mathfrak{s})$ in an auxiliary preunifier ξ that is used by get_factor to instantiate variables in $\mathrm{dom}(\xi)$ that would otherwise belong to $\mathrm{var}(c) \cap U$. This change, together with the embedding outlined in Proposition 7.10 effectively reduces the excess nondeterminacy associated with $\gamma|\mathrm{var}(\mathfrak{s})$. It is also the starting point for extending the unify_2 procedure to handle instantiation systems in which term occurrence cycles are possible.

STEPWISE DERIVATION OF $\mathrm{unification}_2$

The proof of soundness for unify_2 was not entirely satisfying, not only because it re-used a previous proof, but because it failed to invoke the obvious reason, namely that unify_2 a valid implementation of unify_1. The purpose of the following paragraphs is to establish the truth of this fact. Techniques similar to those used below will be needed in Section 9, which relies heavily on meaning-refining transformations of specifications.

---

† This definition of decomposability is equivalent to the one in [SCHM88]. Taken together, the observations in this exercise show that [SCHM88, Theorem 9.1.15] extends to a variety of other systems, including those with strictly typed first-class terms.

Most of the operations in unification$_1$ are *self-implementing*, for various reasons. The notion of *implementation* is extended to syntactic expressions and programming statements. The approach is metamathematical, but fairly elegant, and it extends easily to pointer implementations in Section 9. These ideas allow us to demonstrate that unification$_2$ is a *partial* implementation of unification$_1$, thereby providing an alternate proof of soundness for unification$_2$.

A function f is *self-implementing* iff $g^{\check{}}(f(x)) = f(g^{\check{}}(x))$, for every concrete x.

Exercise 8.6.

A. The multiset operations {E}, size(E), size($\mathcal{S}$), and $\mathcal{S} \cup \mathcal{F}$, are self-implementing, by definition of $g^{\check{}}$.

B. $t\sigma$, $\sigma + \tau$, and $\sigma|U$ are self-implementing, because g is a homomorphism.

C. $u \leq v$, $T_v \subseteq T_u$ are self-implementing, because g is variable-preserving.

D. $t \in T_v$, $T_u \cap T_v$ are self-implementing, because g is type-preserving.

E. $t \notin V$, dom(σ), $E \uplus F$, $U \cap F$, $s \approx t$, $\sigma \approx \tau$, and ($\mathcal{S}$ is merged) are self-implementing, because g is variable-preserving and maps nonvariables to nonvariables.

Exercise 8.7.

A. $\mathcal{S} \setminus \{E\}$ is self-implementing, if E varies over elements of $\mathcal{S}$.

B. Functional composition is self-implementing: $g^{\check{}}(f \circ h) = g^{\check{}}(f) \circ g^{\check{}}(h)$, assuming f and g are abstractable.

C. $var_g(t) \subseteq var(t)$; $g^{\check{}}(base_{\mathcal{S}}) \subseteq base_{g^{\check{}}(\mathcal{S})}$.

D. var(t), cdm(σ), (σ is triangular), and $base_{\mathcal{S}}$ are self-implementing, if g is *regular* in the sense that $g^{\check{}}(var(t)) = var(g(t))$, for all t.

E. A self-implementing function is abstractable, but need not be uniquely implementable.

F. If f is both self-implementing and uniquely implementable, then f_g is just f restricted to concrete values.

For any expression χ, we write $[\![\chi]\!]S$ for the value of χ in the state S. An expression ψ *implements* χ iff $g^{\check{}}([\![\psi]\!]S) = [\![\chi]\!]g^{\check{}}(S)$, for all program states S in which ψ is to be evaluated. In particular, χ is *self-implementing* iff χ implements χ. More generally, ψ *partially implements* χ iff, for all relevant states S, either $g^{\check{}}([\![\psi]\!]S) = [\![\chi]\!]g^{\check{}}(S)$ or $g^{\check{}}([\![\psi]\!]S)$ is an exception and $[\![\chi]\!]g^{\check{}}(S)$ is not an exception.

We regard a programming statement S$ as a specification of a state transformation S#, and in this case, say S# *satisfies* S$. Given a function h from concrete states to abstract states (in this section, $h = g^{\check{}}$), we say S$ *(fully) implements* R$ iff whenever S# satisfies S$, $h^{\check{}}$(S#) satisfies R$. In particular, S$ is *self-implementing* iff S$ implements S$. More generally, S$ *partially implements* R$ iff whenever S# satisfies S$, there exists R# satisfying R$ such that whenever (S, S′) ∈ S# and S′ is successful (i. e., an exception has not been raised), we have $h^{\check{}}$(S, S′) ∈ R#.

Exercise 8.8. Assume z implements y, q implements p, Y implements X, S$ implements R$, S$$_1$ implements R$$_1$, and S$$_2$ implements R$$_2$.

A. (x := z) implements (x := y).

B. (S$$_1$; S$$_2$) implements (R$$_1$; R$$_2$).

C. (**if** q **then** S$$_1$; **else** S$$_2$; **end if**) implements (**if** p **then** R$$_1$; **else** R$$_2$; **end if**).

D. (**loop exit when** q; S$; **end loop**) implements (**loop exit when** p; R$; **end loop**).

E. (**for** x **in** Y **loop** S$; **end loop**) implements (**for** x **in** X **loop** R$; **end loop**),

D. If R$ satisfies the exit condition χ, and ψ implements χ, then S$ satisfies ψ. (A statement S$ satisfies an actual exit condition ψ iff $[\![\psi]\!]$(S×S′), whenever S# satisfies S$ and (S, S′) ∈ S#.)

Exercise 8.9.

A. Assume concrete specifications may contain the symbol **g** denoting the homomorphism g. Then ϕ_g implements ϕ, for any function symbol ϕ.

B. (**pick any** x ∈ Y) implements (**pick any** x ∈ X), if, for all (relevant) S, $[\![Y]\!]S \neq \emptyset$, and $g^{\check{}}([\![Y]\!]S) \subseteq [\![X]\!]g^{\check{}}(S)$.

C. (**select any** q_1 => S$$_1$; ..., q_n => S$$_n$; **end select**) implements (**select any** p_1 => R$$_1$; ..., p_n => R$$_n$; **end select**), if, for all S, $[\![q_1 \vee ... \vee q_n]\!]S$, and, for each i, $[\![q_i]\!]S$ implies $[\![p_i]\!]g^{\check{}}(S)$ and S$$_i$ implements R$$_i$.

D. Assume procedure calls pass parameters using a value–in/value–out semantics. Then one procedure call implements another call, provided each of its "in" parameters implements the corresponding "in" parameter of the other and its body implements that of the other.

Observe that Exercises 8.6 through 8.9 carry over to partial implementations.

Proposition 8.10. The unification$_2$ specification is a partial implementation of the unification$_1$ specification. Consequently, unify$_2$ is sound$_g$. If simplify is deterministic, then unification$_2$ is a full implementation.

Proof. To see that unification$_1$ is implemented by unification$_2$, we first show that unifies_2c$_2$ (the version associated with unification$_2$) implements unifies_2c$_1$ (the version associated with unification$_1$). We need to show that, $g^{\check{}}$(unifies_2c$_2$) satisfies the specification of unifies_2c$_1$, but need only consider concrete states in which var(c) ∩ var(d) = ∅. For these, var($g^{\check{}}$(c)) ∩ var($g^{\check{}}$(d)) = ∅, since g is variable preserving, so that $g^{\sim}$(unifies_2c_2) must satisfy the abstract exit condition $g^{\sim}(\sigma)$ unifies $g^{\sim}(c) \approx g^{\sim}(d)$. But this follows immediately from the exit condition for unifies_2c_2.

We next show that the statement "if $base_{\mathfrak{s}}$ = ∅ then break_cycle($\mathfrak{s}$); end if;" implements the statement "if $base_{\mathfrak{g}}$ = ∅ then raise clash_failure; end if," where $\mathfrak{g} = g^{\sim}(\mathfrak{s})$. If $base_{\mathfrak{s}} \neq \emptyset$, then $base_{\mathfrak{g}} \neq \emptyset$, so that both statements are null. If $base_{\mathfrak{s}} = \emptyset$ and $base_{\mathfrak{g}} \neq \emptyset$, then the abstract statement is null. If break_cycle succeeds, it creates a new state whose image under $g^{\sim}$ is unchanged. Finally, if $base_{\mathfrak{g}} = \emptyset$, then break_cycle cannot succeed and, like its abstract counterpart, raises cycle_failure. This step is only a partial implementation, since break_cycle may fail, even if its abstract counterpart succeeds. However, if break_cycle is deterministic, then its feasibility specification implies that it must succeed, if possible, so that a full implementation is assured.

Finally, by Exercises 8.6 through 8.9, we know all of the other procedures and functions in unification_1 are self-implementing. Consequently, unification_2 partially implements unification_1. Hence, unify_2 is sound_g, by Exercise 8.9A, Theorem 7.4, and Exercise 8.8D. □

TREE-UNIFICATION

One finds a *tree-unifier* for a term system by performing ordinary unification using only those construct unifiers justified by Exercise 7.1B. This is equivalent to unifying a tree-implementation of the term system and then abstracting. Like unification_g, tree-unification is strong enough to give independence of concrete term representations. The main advantage of tree-unification is that construct-unifiers are not an issue, so that first-order unification techniques suffice. Unfortunately, there are also some important disadvantages. In general, not all unifiers are tree-unifiers, and those that are may be harder to find.

If terms are weakly restrictable and no variable is an instance of a nonvariable, and if control nondeterminacy is introduced in the choice of construct factorizations, then unify_1 procedure is tree-complete. Example 8.4 shows that we do need to consider multiple construct factorizations, and this leads to E-graph term implementations in the sense of [NELS80]. With minor modifications including control nondeterminacy in the choice of multiequations to reduce, we can obtain an algorithm that is tree-complete for arbitrary weakly restrictable instantiation systems. The addition of still more control nondeterminacy is justified, in this case, by Example 8.5.

In the special case where $g = |\_| : \mathfrak{B}^t \rightarrow \mathfrak{B}$, we say that an (abstract) substitution σ *tree*-unifies $\mathfrak{s}$ iff there exists ξ and $\mathfrak{X}$ such that $|\xi| = \sigma$, $|\mathfrak{X}| = \mathfrak{s}$, and ξ unifies $\mathfrak{X}$. Clearly, if σ tree-unifies $\mathfrak{s}$, then σ unifies $\mathfrak{s}$. The converse is false (Examples 8.3 and 8.4). A unification procedure is *tree-sound* iff successful executions return only tree-unifiers. It is *tree-complete* iff any tree-unifier is, feasibly, less general than one returned by the procedure.

Proposition 8.11. If no variable is an instance of a nonvariable and $\mathfrak{S}$ is a tree-unifiable merged system, then $\mathfrak{S}$ has a triangular enumeration.

Proof. Suppose $\mathfrak{S}$ has a tree-unifier τ. Let $\mathfrak{X}$, ξ be tree implementations such that ξ unifies $\mathfrak{X}$, $\mathfrak{S} = |\mathfrak{X}|$, and $\tau = |\xi|$. Then $\mathfrak{X}$ is a merged system, since $|\_|$ is variable-preserving, and thus has a triangular enumeration of the form $X_1, \ldots, X_n$, by Proposition 7.2. To see that $|X_1|, \ldots, |X_n|$ is the desired enumeration of $\mathfrak{S}$, suppose $v \in |X_i| \cap \mathrm{var}(|X_j|)$, with $i \leq j$. Then $v = |x|$, for some $x \in X_i$. Moreover, x must be a (uniquely determined) variable, since, otherwise, we would have $v = |x| = |\langle c, x_1, \ldots, x_m\rangle| = c\langle |x_1|, \ldots, |x_m|\rangle$, contradicting the assumption that v is not an instance of a non-variable. Finally, $x \in \mathrm{var}(X_j)$, since $\mathrm{var}(|X_j|) \subseteq |\mathrm{var}(X_j)|^{\sim}$, by Proposition 5.1E. This contradicts our choice of $X_1, \ldots, X_n$. □

Pattern		Replacement
Generic package *unification$_1$*	==>	**Generic package** *unification$_t$*
...		
procedure *restricts*(σ: out; c: in; T: in);		
-- exit ...;		
	==>	-- feasibly, if $c\tau \in T$, then $\sigma \leq \tau$;
procedure *unifies_2c*(σ: out; c, d: in);		
-- exit ... ;	==>	-- exit φ is a renaming from d to c;
		-- termination is feasible;
procedure *factors*(c, σ: out; t in);		
-- entry ... ;	==>	
-- exit ... ;		
	==>	-- feasibly, if $t = d\tau$, then c is a
		-- renaming of d and $\sigma \leq \tau$;
...		
is		
procedure *unify*(δ: out; $\mathfrak{D}$: in);		
-- exit δ unifies $\mathfrak{D}$;	==>	-- exit δ tree-unifies $\mathfrak{D}$;
end unification$_1$;		

Table 8.2: Modification for Tree-Unification

Let $\mathsf{unification}_t$ be obtained from $\mathsf{unification}_1$ by simply respecifying support procedures, as indicated in Table 8.2. Let $\mathsf{unify}_t = \mathsf{unification}_t.\mathsf{unify}$.

Exercise 8.12. Assume, in $\mathfrak{B}$, that substitutions are variable-preserving, that terms are weakly restrictable, and that no variable is an instance of a nonvariable. Then, unify_t is tree-sound and tree-complete.

Exercise 8.13. Drop the assumption that no variable is an instance of a nonvariable. If $\mathfrak{s}$ is tree-unifiable, then it is possible to implicitly construct a unifiable (hence, triangular) tree representation for $\mathfrak{s}$, by modifying the body of $\mathsf{unification}_t$ along the following lines: Introduce a variant of break_cycle in which variables are factored, in order to force new multiequations into $\mathsf{base}_{\mathfrak{s}}$. Keep track of the variable factorizations in an auxiliary function, ξ, and modify vv_reduce, vc_reduce, and get_factor so that they automatically factor variables in the domain of ξ. Replace the "pick any" statement in unify with a corresponding "choose some" control choice. The resulting procedure is tree-sound and tree-complete.

COUNTEREXAMPLES

8.1. *Hidden Concrete Variables.* Let D be an undecidable congruence relation on first-order terms. (McNulty lists several examples of undecidable equational theories [MCNU89].) Let f be a new ternary function symbol; let E be the smallest congruence that contains D and the identity $\mathsf{f(x, y, y) = f(0, y, y)}$. Let $\mathfrak{B}$ be the quotient system induced by E. In this example, let r and s vary over terms that do not contain f.

A. For any variable x, $x \notin var([\mathsf{f}(x, r, s)])$ iff r E s. Thus, in $\mathfrak{B}$, the relation $v \notin var(t)$ is recursively enumerable, but not recursive; the var function is not computable.

B. A multiequation of the form $[x] \approx [\mathsf{f(g}(x), r, s)]$ is unifiable (and has unifier $[\mathsf{0}]/x$) iff $[x] \notin var([\mathsf{f(g}(x), r, s)])$. Thus, unifiability is undecidable in $\mathfrak{B}$.

8.2 *Simple, General Construct Unifiers.* Consider the quotient homomorphism g on first-order term instantiation induced by the following schemas: $\mathsf{f_i(p_i(a_{i+1})) = a_i}$; $\mathsf{g_i(b_i, x) = p_i(f_{i+1}(g_{i+1}(b_{i+1}, x)))}$; $\mathsf{f_i(g_i(b_i, K)) = a_i}$; $\mathsf{g_i(b_i, K) = p_i(a_{i+1})}$. Consider the system $\mathfrak{s} = \{\mathsf{f_1(g_1(b_1, x)) \approx a_1}\}$. If unify_2 has a uniform preference for more-general_g construct unifiers, we get,

	$\mathfrak{s} : \{\mathsf{f_1(g_1(b_1, x)) \approx a_1}\}$	— input
1)	$A : (\mathsf{f_1(u) \approx a_1}$;	$\gamma : \mathsf{p_1(a_2)/u}$.
2)	$\theta : \epsilon$;	$\mathfrak{F} : \{\mathsf{g_1(b_1, x) \approx p_1(a_2)}\}$.
3)	$A : (\mathsf{g_1(v, x) \approx p_1(w)})$;	$\gamma : \mathsf{b_1/v + f_2(g_2(b_2, x))/w}$.
4)	$\theta : \epsilon$;	$\mathfrak{F} : \{\mathsf{b_1 \approx b_1;\ a_2 \approx f_2(g_2(b_2, x))}\}$.

. . . . .

Thus, the algorithm produces a series of a subgoals that are similar to the original problem. But, the algorithm may also choose lighter, less-general construct unifiers, in which case, steps 3 and 4 of the above computation become,

3) A : (g_1(v, x) ≈ p_1(w)); γ : K/x + b_1/v + a_2/w.
4) $\mathcal{F}$: {K ≈ x; b_1 ≈ b_1; a_2 ≈ a_2}
θ : K/x;
...
$\mathcal{E}$: ∅.

Consequently, the hypothesis that unifies_2c(σ, c, d) is able to return σ such that $\sigma \leq_g \leq_{wt} \tau$ [var(c ≈ d)] cannot be weakened to $\sigma \leq_g \tau$ [var(c ≈ d)]. According to Exercise 8.4D, the quotient instantiation system for this example cannot be not well-founded. Indeed, [g_{i+1}(b_{i+1}, x)] properly occurs in [g_i(b_i, x)].

8.3. *Generality of Tree-unification.* Consider the quotient system of first-order term instantiation induced by the equations,

G(x∗y) = H(y∗x),
Q(G(z)) = R(z + 1∗1) = R(z + 2∗2),
Q(H(z)) = R(1∗1 + z) = R(2∗2 + z).

The system $\mathcal{E}$ = {[Q(G(z))] = [Q(H(w))]} has a most general unifier that is not a tree-unifier:

A : ([Q(u)] ≈ [Q(v)]) α : [G(z)]/u + [H(w)]/v
γ : v/u

$\mathcal{E}$: {A} = {v ≈ [G(z)] ≈ [H(w)]}
γ : [G(x∗y)]/v + [(x∗y)]/z + [(y∗x)]/w
. . . .
σ|var($\mathcal{D}$) : [(x∗y)]/z + [(y∗x)]/w. — final result

It is easy to see that this unifier does not lift to any tree implementation of the original term system. However, a tree unifier can be found by choosing tree representations that differ from those implicated in the above unification, namely R(z + 1∗1) for [Q(G(z))] and R(1∗1 + w) for [Q(H(w))]; the resulting tree unifier, 1∗1/z + 1∗1/w, is not most general, even among tree-unifiers.

8.4. *Nondeterminacy and Tree-unification.* Starting with first-order term instantiation, take the quotient system induced by $\mathbf{F(P(x)) = Q(H_1(x)) = Q(H_2(x))}$. Consider the term system $\mathcal{E} = \{[\mathbf{Q(H_1(x))}] \approx [\mathbf{Q(H_2(y))}]\}$. Then $\mathbf{Q}$ is a maximal common construct of the two terms, but the construct reduction achieved by factoring out $\mathbf{Q}$ leaves $\mathcal{F} = \{[\mathbf{H_1(x)}] \approx [\mathbf{H_2(y)}]\}$, which is not unifiable. The successful strategy is to factor out $\mathbf{F}$, obtaining the system $[\mathbf{P(x)}] \approx [\mathbf{P(y)}]$. Consequently, the selection of a particular atomic factorization must be regarded as a control choice when performing tree unification.

8.5. *Tree-unification with Cyclic Terms.* Start with the order-sorted system given by $\mathbf{x, C, D} \in T_1$; $\mathbf{y, A} \in T_2$; $\mathbf{F(y, x)} \in T_1$; $\mathbf{G(x)} \in T_2$. Take the quotient system induced by $\mathbf{x = F(G(C), x)}$; $\mathbf{y = G(F(y, D))}$. Consider the term system $\mathcal{E} = \{[\mathbf{x}] \approx [\mathbf{F(y, C)}]; [\mathbf{y}] \approx [\mathbf{G(x)}]\}$.

A. We can find a tree-unifier, even though this system fails to have a triangular enumeration: Rewrite the first multiequation by factoring $\mathbf{x}$, obtaining $[\mathbf{F(G(C), x)}] \approx [\mathbf{F(y, C)}]$. Then reduce, obtaining a new term system

$$\mathcal{E} = \{[\mathbf{G(C)}] \approx [\mathbf{y}]; [\mathbf{x}] \approx [\mathbf{C}]; [\mathbf{y}] \approx [\mathbf{G(x)}]\}.$$

Merge and reduce twice, obtaining

$$\theta = [\mathbf{G(x)}]/[\mathbf{y}];\ \mathcal{F} = \{[\mathbf{C}] \approx [\mathbf{x}]\};\ \mathcal{E} = \{[\mathbf{x}] \approx [\mathbf{C}]; [\mathbf{x}] \approx [\mathbf{C}]\}$$

Again, merge and reduce twice:

$$\theta = [\mathbf{C}]/[\mathbf{x}];\ \mathcal{F} = \mathcal{E} = \emptyset.$$

B. The above success depended, not only on the choice of what factorization to use, but also on a choice of which multiequation to reduce first: Rewrite the second multiequation as $[\mathbf{G(F(y, D))}] \approx [\mathbf{G(x)}]$ and reduce, obtaining the new system $\mathcal{E} = \{[\mathbf{x}] \approx [\mathbf{F(y, C)}]; [\mathbf{F(y, D)}] \approx [\mathbf{x}]\}$.

Merge, obtaining the multiequation $[\mathbf{x}] \approx [\mathbf{F(y, D)}] \approx [\mathbf{F(y, C)}]$. A straightforward effort at reduction leads to the subgoal $[\mathbf{D}] \approx [\mathbf{C}]$. The only narrowing substitution for $\mathbf{y}$ that allows an identification is $[\mathbf{G(C)}]/\mathbf{y}$, and this gives the subgoal $[\mathbf{x}] \approx [\mathbf{D}] \approx [\mathbf{C}]$, which is clearly not unifiable. Hence, the order in which multiequations are selected must be regarded as a control choice.

SECTION 9
IMPLEMENTATION AND COMPUTATIONAL COMPLEXITY

With the appropriate implementation strategies, nondeterministic N·log(N) execution times can be achieved for a select class of instantiation systems, with two important caveats. We know that control nondeterminacy can introduce infinitely many execution paths (Example 7.2), and that some paths can be infinite as well (Example 7.4). In light of the first problem, we prove only that the time for each particular execution path is bounded by N·log(N), as opposed to the total for all paths. In light of the second problem, we regard control choices as input, and include in N, not only the weight of the term system to be unified, but the weights of all nonvariables introduced by restricts and reduce as well. This relatively weak measure of complexity is still useful, because in typical applications, failure to quickly find a unifier is an indication that the theorem prover is probably working on a wrong subgoal.

We begin with a series of transformational implementation steps designed to ensure efficiency. Every step other than the first is a generalization of strategies used by Martelli and Montanari [MART82, Sections 4 and 6].

1. *Deferred Merging*. The implementations of $base_{\mathcal{B}}$ and merge($\mathcal{B}$) impose an initialization overhead that can be deferred by initially choosing "obvious" base multiequations E such that $E \cap V = \emptyset$. Once a base multiequation is selected, subsequent reductions preserve its status as a base multiequation, and another one needn't be selected until the first one is used up.

2. *Fragment Merging*. After an initial merge, only newly produced fragments fail to be in the merged portion of $\mathcal{B}$. These may be kept in a separate system $\mathcal{F}$, and the resulting merge step, $\mathcal{B}$:= merge($\mathcal{B} \cup \mathcal{F}$), can be implemented with a procedure *merge*($\mathcal{B}$, $\mathcal{F}$) which relies on the fact that $\mathcal{B}$ is merged. We at first define merge in terms of a function $find_{\mathcal{B}}$(v) such that $v \in find_{\mathcal{B}}(v) \in \mathcal{B}$.

3. *Implementation of $find_{\mathcal{B}}$*. This function is replaced with a sparse array *find* that maps variables in $\bigcup\mathcal{B}$ to pointers into $\mathcal{B}$. Pointers are used in order to achieve efficient updating of find.

4. *Implementation of $base_{\mathcal{B}}$*. This function is also implemented as the variable, *base*. New multiequations are added to base during merging, if their *variable dependency count, vdc*(E, $\mathcal{B} \cup \mathcal{F}$), is 0. This step places an additional, mild restriction on the concrete instantiation system $\mathcal{A}$.

5. *Implementation of vdc*(E, $\mathcal{B} \cup \mathcal{F}$). This function is implemented by enhancing the find array with *vdc* fields that are systematically updated to reflect changes in $\mathcal{B} \cup \mathcal{F}$.

6. *Implementation of* δ. The substitution δ is internally implemented as a sequence of substitutions whose composition is δ, so that composition can be deferred and then evaluated from right to left. (Moreover, it may be better to return $\delta|U_0$, where U_0 is the initial estimate of var($\mathfrak{D}$).)
7. *Lower-Level Strategies*. The operations E ∩ V, E \ V, size(E ∩ V) > 1, $\mathfrak{B}$:= $\mathfrak{B}$ ∪ {F}, and (pick E ∈ $\mathfrak{B}$; $\mathfrak{B}$:= $\mathfrak{B}$ \ {E}) can all execute in unit time. Multiequations can be implemented as records with separate fields for variables and nonvariables. Term systems and the fields of a multiequation can be implemented as linked lists (see Exercise 7.7 regarding the ordering of list elements).

Of the above strategies, only the first five are fully worked out. The last two are nonproblematic and are left as exercises for the reader.

The proofs in this section are necessarily less formal than in previous sections, because they involve programming techniques for which no generally accepted formal semantics has been given. Their purpose is to illuminate the presented implementation strategies and to provide informal evidence of correctness.

DEFERRED MERGING

The design nondeterminacy established in Section 8 allows us to select base multiequations in any order. Thus, we may initially pick E ∈ $\mathfrak{B}$ with E ∩ V = ∅, if possible. In this case, E ∈ $base_{\mathfrak{B}}$, and cc_reduce is the only likely reduction. This optimization generalizes a technique suggested by Stillman [STIL73].

Once E is picked, there is no need to pick a different multiequation as long as E contains more than one term. These modifications to the algorithm are presented in Table 9.1, and the fact that they are valid implementation strategies is given in Exercise 9.1.

In the case of a non-unifiable system, deferred merging often allows a clash_failure to be detected before merging and its associated initialization overhead is encountered. Thus, if the statement "pick any U with U ⊇ var($\mathfrak{D}$)" can be implemented in sublinear time, then nonunifiable term systems can often be detected in sublinear time. One possibility for estimating var($\mathfrak{D}$) is to implement terms in such a way that an outer estimate of var(t) is automatically stored with each term t. Another is to use a larger instantiation system for unification than for the term system $\mathfrak{D}$ (as in Proposition 5.5C). In this case, W is just taken to be the set of all variables in the smaller system. If unification is successful, a final step in the algorithm can rename newly introduced variables, in order to push the solution back into the smaller instantiation system, so that variables of the larger system are not visible outside the algorithm.

Let *unify₃* = unification₃.unify, where *unification₃* is given by Table 9.1. Notice that after the first main unify₃ loop, every element of 𝒮 contains a variable.

Exercise 9.1. The unification₃ package given by Table 9.1 is a full implementation of unification₂ (with respect to the identity mapping on program states). Hence, Theorems 8.1 and 8.3 carry over to unify₃.

```
Pattern                               Replacement

Package body unification2       ==>   Package body unification3
   procedure unify ...
   ...
                                ==>     loop
                                           if possible,
                                              pick any E ∈ 𝒮 with E ∩ V = ∅;
                                           else exit end if;
                                           loop exit when size(E) = 1;
                                              cc_reduce(E, 𝒮, U);
                                           end loop;
                                           𝒮 := 𝒮 \ {E};
                                        end loop;
      loop exit exit when 𝒮 = ∅;
         ...
         select any             ==>
            ... =>  ...         ==>   while size(E ∩ V) > 1 loop
                                         vv_reduce(δ, E, 𝒮, U); end loop;
            ... => ...          ==>   if E ∩ V ≠ ∅ and E \ V ≠ ∅ then
                                         vc_reduce(δ, E, 𝒮, U); end if;
            ... => ...          ==>   while size(E \ V) > 1 loop
                                         cc_reduce(E, 𝒮, U); end loop;
            ... => ...          ==>   𝒮 := 𝒮 \ {E};
         end select;            ==>
      end loop;
      ...
   end unify;
   ...
end unification2;               ==>   end unification3;
```

Table 9.1: Deferred Merging

FRAGMENT MERGING

As detailed in the proof of Proposition 9.2, we first split off part of $\mathcal{S}$ into a new "fragment" term system $\mathcal{F}$. New fragments are now added to $\mathcal{F}$ during reduction, base multiequations are still taken from $\mathcal{S}$, and $\mathcal{S}$ and $\mathcal{F}$ are combined whenever merging is performed. This splitting need only be performed in the second of the two main unify loops, since merging is not an issue in the first loop; as a result, we need two slightly different versions of cc_reduce.

This first variable-splitting step causes the tests $\mathcal{S} = \emptyset$ and $\text{base}_{\mathcal{S}} \neq \emptyset$ to be replaced with $\mathcal{S} \cup \mathcal{F} = \emptyset$ and $\text{base}_{\mathcal{S} \cup \mathcal{F}} \neq \emptyset$, respectively. The test $\mathcal{S} \cup \mathcal{F} = \emptyset$ can be simplified to $\mathcal{S} = \emptyset$ by deferring the loop exit test, and the test $\text{base}_{\mathcal{S} \cup \mathcal{F}} \neq \emptyset$, can be simplified to $\text{base}_{\mathcal{S}} \neq \emptyset$ by taking advantage of design nondeterminacy. This is the last step that relies on design nondeterminacy, and we eliminate the "select any" construct at this point. Finally, the assignments ($\mathcal{S}$:= merge($\mathcal{S} \cup \mathcal{F}$); $\mathcal{F}$:= $\emptyset$) are replaced with an incremental merge procedure. The combined effect of these steps is presented in Table 9.2.

If $\mathcal{S}$ is merged, we define $\mathit{find}_{\mathcal{S}}(v)$ to be the unique multiequation E such that $v \in E \in \mathcal{S}$, for each $v \in \bigcup \mathcal{S}$. In Table 9.2, the notation "==+" is used to indicate the creation of a new procedure patterned after an old one, without removal of the old procedure.

Proposition 9.2. The unification_4 specification is a full implementation of the unification_3 specification. Hence, Theorems 8.1 and 8.3 carry over to unify_3.

Proof. The transformation given in Table 9.2 splits into five steps. The second through fifth are marked with corresponding comments in Part II of the table.

1. Split $\mathcal{S}$ into two fragments, as indicated in Part I of Table 9.2. This step clearly preserves the basic structure of the algorithm. It is, in fact, a step-wise implementation with respect to the state abstraction function h given by $h(S).\mathcal{S} = S.\mathcal{S} \cup S.\mathcal{F}$, and $h(S).x = x$, if $x \neq$ "$\mathcal{S}$" and $x \neq$ "$\mathcal{F}$."
2. Defer the loop exit test until after merging and re-initialization of $\mathcal{F}$. (If $\mathcal{S} \cup \mathcal{F} = \emptyset$, then $\text{base}_{\mathcal{S} \cup \mathcal{F}} = \emptyset$, so that the exit test still happens appropriately.)
3. As shown below, $\text{base}_{\mathcal{S}} \subseteq \text{base}_{\mathcal{S} \cup \mathcal{F}}$; design nondeterminacy identified in Section 8 allows us to more restrictively pick $E \in \text{base}_{\mathcal{S}}$, rather than pick $E \in \text{base}_{\mathcal{S} \cup \mathcal{F}}$, if possible, and to (less efficiently) prefer merging before reducing in the case where $\text{base}_{\mathcal{S}} = \emptyset$, but $\text{base}_{\mathcal{S} \cup \mathcal{F}} \neq \emptyset$.
4. Replace the "select any" statement with an "if" statement that avoids merging unless $\text{base}_{\mathcal{S}} = \emptyset$.
5. Replace ($\mathcal{S}$:= merge($\mathcal{S} \cup \mathcal{F}$); $\mathcal{F}$:= $\emptyset$) with merge($\mathcal{S}$, $\mathcal{F}$).

Pattern		Replacement
Package body *unification$_3$*	==>	**Package body** *unification$_4$*
procedure *unify* ...		
is ... $\mathfrak{E}$:= $\mathfrak{D}$;	==>	is ... $\mathfrak{E}$: merged_system := $\emptyset$;
...		$\mathfrak{F}$:= $\mathfrak{D}$; ...
begin		
pick any U ...		
loop ... $\mathfrak{E}$ ... end loop;	==>	loop ... $\mathfrak{F}$ ... end loop;
loop exit when $\mathfrak{E} = \emptyset$;	==>	loop exit when $\mathfrak{E} \cup \mathfrak{F} = \emptyset$;
select any		
$base_{\mathfrak{E}} \neq \emptyset$ => null;	==>	$base_{\mathfrak{E} \cup \mathfrak{F}} \neq \emptyset$ => null;
true =>		
$\mathfrak{E}$:= merge($\mathfrak{E}$);	==>	$\mathfrak{E}$:= merge($\mathfrak{E} \cup \mathfrak{F}$); $\mathfrak{F}$:= $\emptyset$;
if $base_{\mathfrak{E}} = \emptyset$ ... end if;		
end select;		
Pick any E $\in base_{\mathfrak{E}}$;		
... cc_reduce(E, $\mathfrak{E}$, U);	==>	... cc_reduce2(E, $\mathfrak{E}$, $\mathfrak{F}$, U);
... vc_reduce(δ, E, $\mathfrak{E}$, U);	==>	... vc_reduce(δ, E, $\mathfrak{E}$, $\mathfrak{F}$, U);
...		
end loop;		
end unify;		
...		
procedure *cc_reduce*(... $\mathfrak{E}$ ...)	==+	**procedure** *cc_reduce2*(... $\mathfrak{E}$, $\mathfrak{F}$ ...)
...		
$\mathfrak{E} := \mathfrak{E} \cup \{E\} \cup (\gamma \approx \alpha + \beta)$;	==>	$\mathfrak{E} := \mathfrak{E} \cup \{E\}$; $\mathfrak{F} := \mathfrak{F} \cup (\gamma \approx \alpha + \beta)$;
end cc_reduce;		
procedure *vc_reduce*(... $\mathfrak{E}$ ...)	==>	**procedure** *vc_reduce*(... $\mathfrak{E}$, $\mathfrak{F}$ ...)
...		
$\mathfrak{E} := \mathfrak{E} \cup \{E\} \cup (\gamma\|var(c) \approx \alpha)$;	==>	$\mathfrak{E} := \mathfrak{E} \cup \{E\}$; $\mathfrak{F} := \mathfrak{F} \cup (\gamma\|var(c) \approx \alpha)$;
end vc_reduce;		
...		
end unification$_3$	==>	end unification$_4$;

Table 9.2: Fragment Merging (Part I)

Pattern		Replacement	
Package body *unification₄*			
procedure *unify* ...			
is ...			
loop exit when $\mathcal{G} \cup \mathcal{F} = \emptyset$;	==>	loop	-- 2
select any	==>	if	-- 4
$base_{\mathcal{G} \cup \mathcal{F}} \neq \emptyset$ => null;	==>	$base_{\mathcal{G}} = \emptyset$ then	-- 3, 4
true =>	==>		-- 4
$\mathcal{G}$:= merge($\mathcal{G} \cup \mathcal{F}$); $\mathcal{F} := \emptyset$;	==>	merge($\mathcal{G}$, $\mathcal{F}$);	-- 5
		exit when $\mathcal{G} = \emptyset$;	-- 2
if $base_{\mathcal{G}} = \emptyset$ ... end if;			
end select;	==>	end if;	-- 4
pick any $E \in base_{\mathcal{G} \cup \mathcal{F}}$;	==>	pick any $E \in base_{\mathcal{G}}$;	-- 3
...			
end loop;	==>	end loop;	
end unify;			
...			

```
==>  procedure merge(𝒢, ℱ: in out;)                                  -- 5
     is begin
         while ℱ ≠ ∅ loop
             pick any F ∈ ℱ; ℱ := ℱ \ {F}; merge_1(𝒢, F);
         end loop;
     end merge;

==>  procedure merge_1(𝒢: in out; F: in)                              -- 5
     is W := F ∩ V; G := F \ V; H;
     begin
         for v in W loop
             if v ∈ dom(find_𝒢) then
                 H := find_𝒢(v); 𝒢 := 𝒢 \ {H};
             else H := {v}; end if;
             G := G ⊎ H;
         end loop;
         𝒢 := 𝒢 ∪ {G};
     end merge_1;
```

end unification₄

Table 9.2: Fragment Merging (Part II)

To complete the proof, we need to establish the following conditions:

unify invariants: $\text{base}_{\mathcal{E}} \subseteq \text{base}_{\mathcal{E} \cup \mathcal{F}}$; (i)
$\mathcal{E}$ is merged; (ii)

merge exit: $\mathcal{E}' = \text{merge}(\mathcal{E} \cup \mathcal{F})$ and $\mathcal{F}' = \emptyset$;

merge_1 exit: $\mathcal{E}' = \text{merge}(\mathcal{E} \cup \{F\})$.

Initially, and throughout the first main loop, $\mathcal{E} = \text{base}_{\mathcal{E}} = \emptyset$. Merging trivially preserves both loop invariants. If reduction occurs using vv_reduce, a new variable w may be picked from V \ U, but $\text{var}(\mathcal{F}) \subseteq U$, so that E′ will be isolated in $\mathcal{E}' \cup \mathcal{F}$, and $\mathcal{E}'$ is still merged. If reduction occurs using cc_reduce, no new variables will belong to $\cup\mathcal{E}$, and any new variables in $\text{var}(\mathcal{E}' \cup \mathcal{F}')$ are chosen from V \ U, so that E′ is again isolated in $\mathcal{E}' \cup \mathcal{F}'$. Moreover, any other element of $\text{base}_{\mathcal{E}}$ will still be isolated in $\mathcal{E}' \cup \mathcal{F}'$ for the same reason, so that the loop invariants are again preserved. A similar argument shows that the unify invariants are preserved if reduction occurs using vc_reduce.

The correctness of merge follows straightforwardly from Exercise 7.3F and the associativity of multiset union. Regarding the correctness of merge_1, let $\mathcal{H}$ be the multiset of all values of H that result from the assignment ($H := \text{find}_{\mathcal{E}}(v)$). Then, $\mathcal{H} = \{E \in \mathcal{E} \mid E \cap W \neq \emptyset\}$, so that $\mathcal{H} \cup \{F\}$ is a component of $\mathcal{E} \cup \{F\}$. In particular, the multiplicities in $\mathcal{H}$ are correct: in the case where $\text{size}(E \cap W) > 1$, the loop fails to find E a second time, since it has been removed from $\mathcal{E}$, and $\text{find}_{\mathcal{E}}$ returns null. Let W′ be the set of all values of H that result from assignments of the form $H := \{v\}$. Then W′ contains every element of $F \cap V$ that does not belong to $\cup\mathcal{H}$. Consequently, the final value of G is $\uplus(\mathcal{H} \cup \{F\})$, and the merge_1 exit condition follows directly from this. □

IMPLEMENTATION OF find$_{\mathcal{E}}$

We replace $\text{find}_{\mathcal{E}}$ with a sparse array find. When merge_1 merges G and H to form $G \uplus H$, find(v) must be updated. In order to minimize updating, we define find in such a way that it maps variables to access values instead of multiequations. Only variables belonging to the smaller of $G \cap V$ and $H \cap V$ need to be handled. The changes to $\mathcal{E}$ which are made by cc_reduce do not significantly alter the $\text{find}_{\mathcal{E}}$ function, and do not require updating of the find array. Interestingly, the changes to $\mathcal{E}$ which result from removing variables in vv_reduce and vc_reduce also do not require updating of find, despite the fact that $\text{dom}(\text{find}_{\mathcal{E}})$ shrinks, because find(v) is never evaluated for any v that has been removed from $\mathcal{E}$. As an aid in proving this, we accumulate variables removed from $\text{dom}(\text{find}_{\mathcal{E}})$ in the ghost variable Y. Finally, we tentatively update find to reflect changes in $\mathcal{E}$ that occur in break_cycle, even though the following implementation step will make this unnecessary.

With the introduction of access variables, the relation between syntactic objects and the values they denote is more prominent. Access values are essentially global variables that are, themselves, capable of denoting (i.e. pointing to) other values. In the implementation given below, "G" is an access variable and "G.meq" is an associated multiequation variable. Abstractly, both "G" and "G.meq" refer to the same multiequation. As a result, the abstract effect of an assignment of the form G.meq := F depends on which access variables have the same value as G when the statement G.meq := F is executed.

Pattern		Replacement
Package body *unification*$_4$	==>	**Package body** *unification*$_5$
is		
==> **type** *multieq_ptr* = access record meq: multiequation; end record;†		
==> **type** *container* = array (variable) of multieq_ptr;		
procedure *unify* ...		
is ...;	==>	is ...; find: container := (others => null);
		ghost Y := $\emptyset$;
... merge(...);	==>	... merge(find, ...);
... break_cycle(...);	==>	... break_cycle(find, ...);
...		
procedure vv_reduce ...		
... $\delta := \delta\gamma$;	==>	... $\delta := \delta\gamma$; Y := Y ∪ {u, v};
...		
procedure *vc_reduce* ...		
... $\delta := \delta(\gamma \vert \{v\})$;	==>	... $\delta := \delta(\gamma \vert \{v\})$; Y := Y ∪ {v};
...		
procedure *break_cycle*(...)	==>	**procedure** *break_cycle*(find: in out;...)
is ...;	==>	is G: multieq_ptr;
... $\mathfrak{g} := \mathfrak{g} \setminus \{F\} \cup \{E\}$;		
	==>	G := new multieq_ptr(E);
		for v in E loop find(v) := G; end loop;
...		
procedure *merge*(...)	==>	**procedure** *merge*(find: in out; ...)
... merge_1(...);	==>	... merge_1(find, ...);
...		

Table 9.3: Implementation of find$_{\mathfrak{g}}$

---

† A one-element record is used here in order to facilitate the following implementation step.

```
Pattern                                  Replacement
procedure merge_1(...)          ==> procedure merge_1(find: in out; ...)
is ... G := F \ V; H;           ==> is ... G := new multieq_ptr(F \ V);
                                          H, H2: multieq_ptr;
begin
   for v in W loop
      if v ∈ dom(find_𝔤) then    ==>  H := find(v);
         H := find_𝔤(v);         ==>  if H ≠ G then
                                          if H ≠ null then
         𝔤 := 𝔤 \ {H};           ==>           𝔤 := 𝔤 \ {H.meq};
                                             if size(G.meq ∩ V) <
                                                size(H.meq ∩ V) then
                                                H2 := G; G := H; H := H2;
                                             end if;
      else H := {v}; end if;     ==>       else H := new multieq_ptr({v}); end if;
      G := G ⊎ H;                ==>       G.meq := G.meq ∪ H.meq;
                                 ==>       for w in H.meq ∩ V loop find(w) := G;
                                          end loop;
                                      end if;
   end loop;
   𝔤 := 𝔤 ∪ {G};                ==> 𝔤 := 𝔤 ∪ {G.meq};
end merge_1;
end unification4;               ==> end unification5;
```

Table 9.3: Implementation of find$_{\mathfrak{g}}$ (Concluded)

We associate each type T with an infinite set *access* T of scalar *access* values that is disjoint from previous sets of scalar values, including boolean, integer, and so forth. The *null* access value belongs to every access type. If T and T′ are distinct types, then (access T) ∩ (access T′) = {null}. The runtime state of a program using access values contains a distinguished component, *heap* that is a partial function defined on access values. If v is a variable of type access T, then the compound variable v.all is synonymous with heap(v); in other words, the value of v.all in state S is (S.heap)(S.v), assuming this is defined. If T is a record type containing the field meq, for example, then v.meq abbreviates v.all.meq. By convention, the null access value never belongs to the domain of heap, so that null.all is always undefined.

New access values may be added to the heap via statements of the form v := new access T (x), where x is of type T and v is of type access T. This statement satisfies the exit conditions v′ ∉ dom(heap) ∪ {null}, dom(heap′) = dom(heap) ∪ {v′}, and v′.all′ = x, where v′.all′ = heap′(v′).

We now introduce an overloaded family of *abstraction* functions h, that eliminate access values from concrete data structures. They are defined as follows, where S is a concrete state and G is of type multieq_ptr, as defined in Table 9.3:

values not involving pointers:	h(x, S) = x.
multieq_ptr:	h(G, S) = S.heap(G).meq.
container:	h(find, S)(v) = S.heap(find(v))
states:	h(S).v = h(S.v, S), provided v is a component of the abstract state.
state transformations:	$h^{\sim}(S\#) = \{\langle h(S), h(S')\rangle \mid \langle S, S'\rangle \in S\#\}$.

We carry over the notion of implementation for statements given in Section 8. Thus, S$ implements R$ iff whenever S# satisfies S$, $h^{\sim}(S\#)$ satisfies R$. For expressions, we must account for the fact that abstraction is state-dependent: ψ *fully implements* χ iff $h([\![\psi]\!]S, S) = [\![\chi]\!]h(S)$, for all states S in which ψ is to be evaluated. Exercise 8.8 straightforwardly carries over to the present situation, showing that most statement constructs are implemented without difficulty.

Exercise 9.3. Assume E is a multiequation in both concrete and abstract states; assume, in concrete staates, that G, H are of type multieq_ptr, and find is of type container; assume, in abstract states, that G is a multiequation.

A. Any expression that does not involve access values is self implementing.
B. G and **G.meq** both implement G.
C. (G := H) implements (G := H).
D. (G := **new multieq_ptr** (E)) also implements (G := E).
E. (**G.meq** := E) implements (G := E), if executed in a state where no access variable other than G has the same value as G.
F. Consider only concrete states and variables u such that,

   $u \notin Y$;
   $u \in U_{\mathcal{E}}$ iff find(u) $\neq$ null;
   if $u \in U_{\mathcal{E}}$, then $u \in$ find(u).meq $\in \mathcal{E}$.

   Then

   find(u) $\neq$ **null** implements $u \in$ **dom**(**find**$_{\mathcal{E}}$);
   find(u) implements **find**$_{\mathcal{E}}$(u) in states where $u \in$ dom(find$_{\mathcal{E}}$).

Proposition 9.4. The procedure *unify*$_5$ = unification$_5$.unify is a full implementation of unify$_4$. Hence, Theorems 8.1 and 8.3 carry over to unify$_5$.

Proof. The changes to procedures other than unify and merge_1 are superficial at most. Moreover, it should also be clear that unify$_5$ is correctly implemented, provided it properly maintains (the abstract image of) find. We first prove merge_1 is correctly implemented, assuming the following concrete assertions, for all values of u:

unify invariant: $Y \supseteq dom(\delta)$ and $Y \cap var(\mathcal{E} \cup \mathcal{F}) = \emptyset$;

merge_1 loop inv: $u \in \cup(\mathcal{E} \cup \{G.meq\})$ iff $find(u) \neq null$ and $u \notin Y$; (i)
if $u \in \cup(\mathcal{E} \cup \{G.meq\})$, then
$u \in find(u).meq \in \mathcal{E} \cup \{G.meq\}$; (ii)
$G.meq \cap E \cap V = \emptyset$, for all $E \in \mathcal{E}$; (iii)

merge_1 exit: $u \in \cup\mathcal{E}'$ iff $find'(u) \neq null$ and $u \notin Y$;
if $u \in \cup\mathcal{E}'$, then $u \in find'(u).meq \in \mathcal{E}'$.

Observe that H is used only in the body of the merge_1 loop and is written before being read; we may therefore treat H as being local to the body of the loop. We consider three cases. In the first, $H \neq G$ and either find(v) = null, or else $size(G.meq \cap V) \geq size(H.meq \cap V)$. For this case, implementation may be justified as follows:

```
for v in W loop
    H := find(v);                                   — Ex 9.3F, 9.3C, (ii), H ≠ G
    if H ≠ G then                                   — assumed true
        if H ≠ null then                            — v ∉ Y, Ex 9.3F, (i), H ≠ G
            𝓔 := 𝓔 \ {H.meq};                       — Ex 9.3A, 9.3B
            if ... end if;
        else H := new multieq_ptr({v}); end if;     — Ex 9.3D
        G.meq := G.meq ∪ H.meq;                     — (see below).
        for ... loop ... end loop
    end if;
end loop;
```

In the second case, where $size(G.meq \cap V) < size(H.meq \cap V)$, G and H are swapped, but the resulting value of G.meq is the same. Consequently, the loop body has the same abstract effect, since H is effectively local to the loop body.

Finally, in the case where find(v) = H = G, v belongs to a multiequation that has already been deleted from $\mathcal{E}$ and merged into G; in this case, the abstract algorithm redundantly merged {v} into G, whereas the new algorithm does nothing. This fact, together with invariant (iii), shows that, in implementing $G := G \uplus H$, it is acceptable to replace $\uplus$ with $\cup$, because $G.meq \cap H.meq = \emptyset$. The assignment to G.meq correctly implements assignment to G, by Exercise 9.3E, since the values of **G** and **H** both result from calls to new.

The unify invariant regarding Y may be briefly argued as follows: since Y is updated whenever dom(δ) grows, we clearly have $Y \supseteq dom(\delta)$. New variables added to Y are also added to U, and any variables added to $\mathcal{E}$ or $\mathcal{F}$ later will be taken from $V \setminus U$ and so will not belong to Y.

We next prove the merge_1 loop invariant and exit conditions, assuming the following:

merge_1 entry: $u \in U\mathcal{S}$ iff find(u) $\neq$ null and $u \notin Y$;
if $u \in U\mathcal{S}$, then $u \in$ find(u).meq $\in \mathcal{S}$;
$Y \cap var(\mathcal{F}) = \emptyset$.

The merge_1 loop invariants hold on entry to the merge_1 loop. In particular, invariants (i) and (ii) follow directly from merge_1 entry. The induction step is also straightforward. In particular, the propagation of invariant (iii) rests on the fact that $\mathcal{S}$ is a merged system. The merge_1 exit conditions follow directly from the merge_1 loop invariants.

Finally, we show that unify honors the merge_1 entry conditions. The third entry condition follows directly from the unify invariant. The first two merge_1 entry conditions hold trivially after initialization and are unaffected by the first unify loop. We now know the first two are preserved across the initial merge in $unify_5$. They are also preserved in the second unify loop, because any variables removed from $U\mathcal{S}$ during reduction are simultaneously added to the ghost variable Y. □

IMPLEMENTATION OF $base_{\mathcal{S}}$

The $base_{\mathcal{S}}$ function is implemented as the term-system *base*. The concrete algorithm decides when to add new multiequations to base by means of a *variable dependency count*, *vdc*(E, $\mathcal{F}$), that counts the number of other occurrences in $U\mathcal{S}$ of variables in $\mathcal{F}$. Thus, $E \in base_{\mathcal{F}}$ iff $E \in \mathcal{F}$ and vdc(E, $\mathcal{F}$) = 0.

We avoid storing multiple copies of multiequations by representing $\mathcal{S}$ as a set $\mathcal{J}$ of access values, with $\mathcal{J}$ containing the range of the find array; base is also represented as a set of access values, for the same reason. ($\mathcal{J}$ and base are sets in the sense that no access value occurs more than once, despite the use of multiset operations.)

Let voc(E, $\mathcal{F}$) = Σ\{voc(v, t) | $v \in E$ and $t \in U\mathcal{F}$\}; let vdc(E, $\mathcal{F}$) = voc(E, $\mathcal{F}$) − size($E \cap V$). (Usually, vdc(E, $\mathcal{F}$) $\geq$ 0, because $E \in \mathcal{F}$.) By definition, $E \in base_{\mathcal{F}}$ iff E is isolated in $\mathcal{F}$ and E precedes every element of $\mathcal{F}$, which happens iff $E \in \mathcal{F}$ and vdc(E, $\mathcal{F}$) = 0.

For convenience of discussion, we add a ghost variable bşe to the abstract state representing the abstract interpretation of the new base variable. An overloaded abstraction function h is given implicitly by the following identities which relate abstract state components to corresponding concrete components and show how to extract abstract values from the concrete state:

$[\![\mathbf{E}]\!]h(S) = h([\![\mathbf{J}]\!]S, S) = [\![\mathsf{J.meq}]\!]S.$
$[\![\mathfrak{S}]\!]h(S) = h([\![\mathfrak{J}]\!]S, S) = [\![\{\mathsf{J.meq} \mid \mathsf{J} \in \mathfrak{J}\}]\!]S.$
$[\![\mathbf{bse}]\!]h(S) = h([\![\mathbf{base}]\!]S, S) = [\![\{\mathsf{J.meq} \mid \mathsf{J} \in \mathbf{base}\}]\!]S.$
$[\![\mathsf{x}]\!]h(S) = h([\![\mathsf{x}]\!]S, S) = [\![\mathsf{x}]\!]S$, for other components x.

Exercise 9.5. Assume F is a multiequation (in both abstract and concrete states), H is of type multieq_ptr and implements the multiequation G, $\mathfrak{g}$ is of type internal_system (see Table 9.4) and implements the term system $\mathfrak{F}$.
A. (**J** := H) implements (**E** := G).
B. ($\mathfrak{J}$:= $\mathfrak{g}$) implements ($\mathfrak{S}$:= $\mathfrak{F}$).
C. (**J.meq** := F) implements ($\mathfrak{S}$:= $\mathfrak{S}$ \ {E}; **E** := F; $\mathfrak{S}$:= $\mathfrak{S}$ ∪ {F}), when executed in states where J ∈ $\mathfrak{J}$ \ base.
D. (**J.meq** := F) implements
($\mathfrak{S}$:= $\mathfrak{S}$ \ {E}; **bse** := **bse** \ {E}; **E** := F; **bse** := **bse** ∪ {F}; $\mathfrak{S}$:= $\mathfrak{S}$ ∪ {F}), when executed in states where J ∈ $\mathfrak{J}$ ∩ base.

Let $unify_6$ = $\mathsf{unification_6.unify}$, where $\mathsf{unification_6}$ is given by Table 9.4.

Proposition 9.6. Assume variable-occurrence counts are well-defined. Then $unify_6$ is a full, stepwise implementation of $\mathsf{unify_5}$. Consequently, those portions of Theorems 8.1 and 8.3 relating to $\mathsf{unify_2}$ carry over to $\mathsf{unify_6}$.

Proof. The variables E and $\mathfrak{S}$ are correctly implemented, directly as a consequence of Exercise 9.5. If $\mathrm{base}_{\mathfrak{S}}$ is implemented correctly, then it it follows straightforwardly that $\mathsf{unify_6}$ is a stepwise implementation of $\mathsf{unify_5}$. (The **break_cycle** procedure requires some minor additional justification which is left as an exercise for the reader.) We need to prove that the following concrete invariants hold the **unify** and **merge** procedures, where $\mathfrak{S}$ abbreviates {J.meq | J ∈ $\mathfrak{J}$} and bse abbreviates {J.meq | J ∈ base}:

$$\text{merge inv:}\quad \mathsf{bse} = \mathrm{base}_{\mathfrak{S}} \cap \mathrm{base}_{\mathfrak{S} \cup \mathfrak{F}}.$$

$$\text{unify loop2 inv:}\quad \mathrm{base}_{\mathfrak{S}} \subseteq \mathrm{base}_{\mathfrak{S} \cup \mathfrak{F}}; \qquad \text{(i)}$$
$$\mathfrak{S} \text{ is merged}; \qquad \text{(ii)}$$
$$\mathsf{bse} = \mathrm{base}_{\mathfrak{S}}. \qquad \text{(iii)}$$

To prove that **merge** maintains its invariant, we notice first that the invariant holds on entry: it does for the initial call, where bse = ∅, and each {v} ∈ $\mathfrak{S}$ fails to precede some element of $\mathfrak{F}$. It also holds for calls inside the second unify loop, as a result of the **unify** loop2 invariants (i) and (iii). To see that it is preserved by **merge_1**, notice, first, that each E ∈ **bse** which is not removed from $\mathfrak{S}$ remains isolated in $\mathfrak{S} \cup \mathfrak{F}$, by Exercise 7.3H, so that E continues to belong to $\mathrm{base}_{\mathfrak{S}}$ and $\mathrm{base}_{\mathfrak{S} \cup \mathfrak{F}}$, as well as **bse**. The new element of $\mathfrak{S}$, G.meq, is not added to **bse** unless vdc(G.meq, $\mathfrak{S} \cup \mathfrak{F}$) = 0, which happens iff it belongs to both $\mathrm{base}_{\mathfrak{S}}$ and $\mathrm{base}_{\mathfrak{S} \cup \mathfrak{F}}$.

```
Pattern                                       Replacement

Package body unification_5           ==> Package body unification_6

   ...
                         ==>   type internal_system = multiset of multieq_ptr;

procedure unify ...
is ... 𝒮: ... := ∅; ... E; ...       ==>   is ...   E; J: multieq_ptr; ...
                                                𝒥, base: internal_system := ∅; ...
begin
   ...
   loop
      if base_𝒮 = ∅ then             ==>   if base = ∅ then
         merge(find, 𝒮, ℱ); ...      ==>      merge(base, find, 𝒥, ℱ); ...
         if base_𝒮 = ∅ then          ==>      if base = ∅ then
            break_cycle(find, 𝒮, U); ==>         break_cycle(base, find, 𝒥, U);
         end if;
      end if;
      pick any E ∈ base_𝒮;           ==>   pick any J ∈ base;
      ... vv_reduce(... E, 𝒮 ...);   ==>   ... vv_reduce(... J.meq  ...);
      ... vc_reduce(... E, 𝒮 ...);   ==>   ... vc_reduce(... J.meq  ...);
      ... cc_reduce2(E, 𝒮 ...);      ==>   ... cc_reduce2(J.meq  ...);
      𝒮 := 𝒮 \ {E};                  ==>   𝒥 := 𝒥 \ {J}; base := base \ {J};
            ... E ...                ==>   ... J.meq ...
   end loop;
end unify;

procedure vv_reduce(... 𝒮, ...)     ==> procedure vv_reduce(... ...)
   ... 𝒮 := 𝒮 \ {E};                 ==>
   ... 𝒮 := 𝒮 U {E};                 ==>
   ...

procedure cc_reduce2(... 𝒮, ...)    ==> procedure cc_reduce2(... ...)
   ... 𝒮 := 𝒮 \ {E};                 ==>
   ... 𝒮 := 𝒮 U {E};                 ==>
end cc_reduce;

procedure vc_reduce(... 𝒮, ...)     ==> procedure vc_reduce(... ...)
   ... 𝒮 := 𝒮 \ {E};                 ==>
   ... 𝒮 := 𝒮 U {E};                 ==>
end vc_reduce;
```

Table 9.4: Implementation of $base_{\mathcal{S}}$

```
Pattern                                   Replacement

procedure break_cycle(...)           ==> procedure break_cycle(base, find, ...)
is s; E; ℱ : constant := ℰ; ...      ==> is s; F; ...
begin
   for F in ℱ loop  E := F;          ==>    for G in 𝒢 loop  F := G.meq;
      for t in F \ V loop
         s := t; simplify(s);
         E := E \ {t} ∪ {s};         ==>          G.meq := G.meq \ {t} ∪ {s};
      end loop;
      ℰ := ℰ \ {F} ∪ {E};            ==>
      G := new multieq_ptr(E);       ==>
      for v in E loop ... end loop;  ==>
   end loop;
   if base_ℰ = ∅ then ... end if;    ==>    init_base(base, find, 𝒢);
                                            if base = ∅ then ... end if;
end break_cycle;

procedure merge(... ℰ ...)           ==> procedure merge(base: in out; ... 𝒢 ...)
   ... merge_1(... ℰ ...); ...       ==>    ... merge_1(base, ℱ, ... 𝒢 ...); ...

procedure merge_1(... ℰ ...)         ==> procedure merge_1(base: in out; ℱ: in;
                                                                ... 𝒢 ...)
is ...
begin
   ... ℰ := ℰ \ {H.meq}; ...         ==>    ... 𝒢 := 𝒢 \ {H}; ...
   ... ℰ := ℰ ∪ {G.meq};             ==>    ... 𝒢 := 𝒢 ∪ {G};
                                               if vdc(G.meq, ℰ ∪ ℱ) = 0 then
                                                  base := base ∪ {G}; end if;
end merge_1;

      ==>   Procedure init_base(base: out := ∅; 𝒢: in out)
            is begin
               for J in 𝒢 loop
                  If vdc(J.meq, ℰ) = 0 then base := base ∪ {J}; end if;
               end loop;
            end init_base;
```

Table 9.4: Implementation of $base_{\mathcal{E}}$ (Concluded)

The first two unify loop2 invariants come from the proof of Proposition 9.2, and will remain true, pending proof of correct implementation. The third invariant is established as follows: it holds on entry to the second loop, as a result of the merge invariant applied on exit from merge, and it is preserved by merge for the same reason. It is preserved by break_cycle, because it is directly re-established by init_base. Finally, it is preserved by each reduction, since the selected multiequation is eventually removed from both $\mathcal{S}$ and bse. (In fact, it is preserved during the reduction process itself, by Exercises 9.5C and 9.5D.) □

IMPLEMENTATION OF vdc(E, $\mathcal{S}$)

We add a new *vdc* field to the find array, in such a way that, for each $E \in \mathcal{S}$, either $E \cap V = \emptyset$ (so that vdc(E, $\mathcal{S}$) = 0), or else E = find(v).meq, for some $v \in E$ and, in this case, vdc(E, $\mathcal{S}$) is stored in the new field, find(v).vdc. These vdc fields are initialized just before the second unify loop and are systematically updated during subsequent reduction and merging. Updating of the vdc fields is simplified by padding $\mathcal{S}$ with trivial multiequations of the form {v}, so that the algorithm can uniformly track vdc(v, $\mathcal{S}$), for all $v \in$ var($\mathcal{S}$).

Updating of the vdc fields is based on additivity properties given in Exercise 9.7. In general, re-initialization of all vdc fields is necessary after a call to break_cycle.

Exercise 9.7. Assume variable-occurrence counts are well-defined.

A. If $E \cap F \cap V = \emptyset$, then vdc($E \cup F$, $\mathcal{S}$) = vdc(E, $\mathcal{S}$) + vdc(F, $\mathcal{S}$).

B. vdc(E, $\mathcal{S} \cup \mathcal{F}$) = vdc(E, $\mathcal{S}$) + vdc(E, $\mathcal{F}$).

C. If $c\alpha = t$, then vdc(E, t)) = vdc(E, $\{v\alpha \mid v \in \text{var}(c)\}$).

Let $unify_7$ = unification_7.unify, where unification_7 is given by Table 9.5.

Proposition 9.8. Assume variable-occurrence counts are well-defined. Then $unify_7$ is a step-wise implementation of unify_6, except for the occasional addition of trivial multiequations of the form {v}. Consequently, those portions of Theorems 8.1 and 8.3 relating to unify_2 carry over to unify_7.

Proof. As in previous proof, we continue to use $\mathcal{S}$ as an abbreviation for $\{J.\text{meq} \mid J \in \mathcal{J}\}$. To show that the vdc fields are correctly updated during merging and reduction, we need to show that the following invariants hold throughout the unify and merge procedures:

$$\text{var}(\mathcal{S} \cup \mathcal{F}) \subseteq \bigcup\mathcal{S}; \qquad \text{(i)}$$

$$\text{vdc}(\text{find}(v).\text{meq}, \mathcal{S} \cup \mathcal{F}) = \text{find}(v).\text{vdc}. \qquad \text{(ii)}$$

The first invariant is needed to facilitate computation of variable dependency counts. It also ensures that the test $H \neq$ null in merge_1 always succeeds, so that it can be eliminated.

```
               Pattern                                       Replacement

Package body unification6 is           ==> Package body unification7 is
   type multieq_ptr = access
      record meq: multiequation;
                                        ==>          vdc: natural;
      end record;
    ...
procedure unify ...
is ...
                                        ==>          init_vocs(find, 𝒢);
   loop ... end loop;
end unify;
procedure cc_reduce2(...)               ==> procedure cc_reduce2(find, ...)
is  ...
   ... 𝓕 := 𝓕 U (γ ≈ α + β);          ==>       if size(E) > 1 then
                                                     add_vocs(find, 𝒢, cγ); end if;
                                                 for v ∈ var(c ≈ d) loop
                                                     add_vocs(find, 𝒢, vγ);
                                                     𝓕 := 𝓕 U {vγ ≈ v(α + β)};
                                                 end loop;
end cc_reduce;
procedure vc_reduce(...)                ==> procedure vc_reduce(find, ...)
is  ...
   ... 𝓕 := 𝓕 U (γ|var(c) ≈ α);        ==>       if size(E) > 1 then
                                                     add_vocs(find, 𝒢, cγ); end if;
                                                 for v ∈ var(c) loop
                                                     add_vocs(find, 𝒢, vγ);
                                                     𝓕 := 𝓕 U {vγ ≈ vα};
                                                 end loop;
end vc_reduce;
procedure break_cycle                   ==> procedure break_cycle
   ... for G in 𝒢 loop ...
                                        ==>          G.vdc := 0;
      end loop;
                                        ==>      init_vocs(find, 𝒢);
   ...
procedure merge                         ==> procedure merge
   ... merge_1(... 𝓕, ...);             ==>      ... merge_1( ... ...);
   ...
```

Table 9.5: Implementation of vdc(E, 𝒢)

```
            Pattern                                   Replacement

procedure merge_1(... 𝓕: in; ...) ==> procedure merge_1(... ...)
is ... G := new multieq_ptr(...); ...==> is ... G := new multieq_ptr(..., 0);
begin
    ...
    if H ≠ G then
    ... if H ≠ null then              ==>
            ...
        else ... end if;              ==>
    ... G.meq := G.meq ∪ H.meq;
                                      ==>          G.vdc := G.vdc + H.vdc - 1;
        ...
    end if;                           ==>       else G.vdc := G.vdc - 1; end if;
    ...
    if vdc(G.meq, 𝓑 ∪ 𝓕) = 0 then==>       if G.vdc = 0 then
    ...

Procedure init_base ...
is
begin
    ...
        If vdc(J.meq, 𝓑) = 0 then   ...  ==>    If J.meq ∩ V = ∅ then
                                                     base := base ∪ {J};
                                                 else pick any v ∈ J.meq;
                                                     if find(v).vdc = 0 then
                                                         base := base ∪ {J}; end if;
                                                     end if;
        end if;
    ...
end init_base;

        ==>     Procedure init_vocs(find: in out; 𝓙: in out)
                is 𝔍 : const := 𝓙;
                begin
                    for J in 𝔍 loop
                        for t in J.meq \ V loop add_vocs(find, 𝓙, t); end loop;
                    end loop;
                end init_vocs;
```

Table 9.5: Implementation of vdc(E, 𝓑) (Continued)

Pattern **Replacement**

```
==>   Procedure add_vocs(find, 𝒮: in out; t)
      is c; α;
      begin
         if t ∈ V then
            if find(t) = null then find(t) := new multieq_ptr({t}, 1);
               𝒮 := 𝒮 ∪ {find(t)};
            else find(t).vdc := find(t).vdc + 1; end if;
         else factors(c, α, t);
            for v in var(c) loop add_vocs(find, 𝒮, vα); end loop;
         end if;
      end add_vocs;
end unification6                       ==> end unification7;
```

Table 9.5: Implementation of vdc(E, 𝒮) (Concluded)

The first invariant is established by init_vocs, by adding {v} to an initially empty 𝒮, whenever v ∈ var(𝒮). It is preserved because only variables from base multiequations can be removed from ∪𝒮, and because any new nonbase variables are encountered during reduction by additional calls to add_vocs.

The second invariant is also established by init_vocs. This initialization step requires, on entry, that find(v).vdc = 0, whenever find(v) ≠ null, a condition that holds both before the second unify loop and after the outer loop in break_cycle. The init_vocs procedure relies on the add_vocs procedure for the following exit conditions (whose verification is omitted):

if v ∈ dom(find), then find(v).vdc′ = find(v).vdc + voc(v, t).
if v ∉ dom(find), then find(v).vdc′ = voc(v, t).

The init_vocs procedure thus associates with each find(v).vdc, the number Σ\{voc(u, t) | u ∈ find(v).meq, t ∈ ∪𝒮 \ V\}; this number coincides with vdc(find(v).meq, 𝒮), because 𝒮 is merged and contains find(v).meq.

The second invariant is preserved by merge_1, by Exercises 9.7A, and the fact that the appropriate count is decremented whenever a variable occurrence is removed. To see that the vdc fields are correctly updated during reduction, first notice that the value of 𝒮 at the top of the second main unify loop never contains a multiequation of the form {t}, where t is a nonvariable: this property holds on entry to the second loop, because such trivial multiequations are filtered out during the first loop, and it is clearly preserved by merging, and by the reduction and deletion of a multiequation. In order to avoid unnecessary computation when a multiequation is deleted, we treat its deletion together with the final updating of the vdc fields. During vv_reduce, no updating is necessary, because the relevant terms, namely u, v, and w, occur only in E ∩ V. During vc_reduce, the nonvariable t is removed from 𝔈 and replaced by {vα | v ∈ var(c)}, which preserves the variable dependency counts, by Exercise 9.7C. The multiset {vγ | v ∈ var(c)} is added, and the appropriate updates are accomplished by add_vocs. Finally, either {cγ} is added, and it is then either immediately removed, or else the vdc fields are updated. Finally, we omit the similar verification of the fact that cc_reduce also updates the vdc fields correctly. □

FURTHER OPTIMIZATIONS

It is possible to further develop the present algorithm to obtain the well-known algorithms of Jaffar and of Martelli & Montanari as special cases. However, it may be more productive to simply list these and related ideas as informal recommendations and conjectures.

Minor Optimizations. Many of the procedures in unification_7 occur only once and should be marked for in-line expansion. The support procedures were deliberately not optimized, in order to facilitate easy use of the algorithm. For example, the statement pair (unifies_2c(γ, c, d); rename_cdm(γ, var(c ≈ d), U)) could be expanded and reimplemented as a single action.

Use of break_cycle. Theorem 8.3 suggests that it is (in principle) only necessary to apply break_cycle to the current term system once a cycle has been detected, but this application implicitly assumes that preceding reduction steps have made a series of fortuitous choices which preserve the eliminability of variables in var(𝔇) \ var_g(𝔇). Consequently, it may actually be better to apply simplify at the beginning of the unification algorithm. In many applications, simplify will be a deterministic algorithm that applies variable-eliminating rewrite rules. In this case, break_cycle need not be called more than once during the course of a unification, so that it may be preferable to unify only term systems that have already been simplified, in which case, break_cycle can be removed from the unification algorithm.

Elimination of 𝔖. If break_cycle is applied directly to 𝔇 or eliminated entirely, then the only remaining use of 𝔖 is in the exit test 𝔖 = ∅ in the second unify loop. In this case, 𝔖 may be replaced with an integer counter, count, that is updated in such a way as to maintain the invariant count = size(𝔖), where 𝔖 is now a ghost variable.

Elimination of Nondeterminacy. As described in Exercise 8.5, control nondeterminacy can be eliminated by restricting attention to systems with finite type intersections, uniformly restrictable constructs, decomposable terms, and unique quotients.

Common-Part Reduction. For this restricted class of systems, a unifiable multiequation is of the form ... $u \approx c\alpha_1 \approx \ldots \approx c\alpha_n$; reduction can produce a partial unifier of the form $c\eta/u$ and a fragment system of the form $\eta \approx \alpha_1 \approx \ldots \approx \alpha_n$, where η is a weakening for c. We consider each $v_i \in var(c)$. For a given i, if $v_i\alpha_j \notin V$, for all j, then $v_i\eta \approx v_i\alpha_1 \approx \ldots \approx v_i\alpha_n$ is a new base multiequation that can be reduced immediately; this observation may be applied recursively to obtain a "common-part" unifier of the form $t_i/v_i\eta$. For those i where $v_i\alpha_j \in V$, for some j, we just let $t_i = v_i\eta$. Finally, the partial unifiers can be composed with the original unifier, $c\eta/u$, to obtain a common-part unifier $c\langle t_1, \ldots, t_n\rangle/u$ for the original multiequation. With this optimization, we get what is essentially the unification algorithm of [MART82].

Deferred Recognition of Base Multiequations. If nondeterminacy is eliminated, the only other reason for preferring base multiequations is the retyping problem illustrated in Example 7.6. If we restrict attention to systems that are strictly typed or are nearly so, then the retyping problem is not an issue, and all reductions may be performed using pure fragmentation (in the sense of Example 7.5), without bothering to compute variable-dependency counts. This observation is the main distinguishing feature of Jaffar's Algorithm [JAFF84].

COMPUTATIONAL COMPLEXITY

The nonlinear complexity of the unification algorithm stems from the implementation of the merge function, a fact which can be seen already in unification_5. The complexity analysis decomposes the unification algorithm into time spent in processing constructs and time spent in processing variables. Construct weights (as defined in Proposition 6.9) are used to show that construct processing is linear in N, provided N includes not only the weight of the term system to be unified, but nonvariables returned by restricts and unify as well. This allows us to conclude that time for variable-processing is $N \cdot \log(N)$.

Exercise 9.9. Assume that atomic operations in unification_5.merge execute in unit time.

A. Pick $n > 0$, define $\mathcal{F}_n$ as follows. For each i with $1 \leq i \leq n$ and each j with $0 \leq j \leq 2^{n-i} - 1$, let F_{ij} be the multiequation $v_p \approx v_q$, where $p = j \cdot 2^i$ and $q = p + 2^{i-1}$. Let $\mathcal{F}_n = \{F_{ij} \mid 1 \leq i \leq n \text{ and } 0 \leq j \leq 2^{n-1} - 1\}$. Thus $\mathcal{F}_4$, for example, may be depicted as follows:

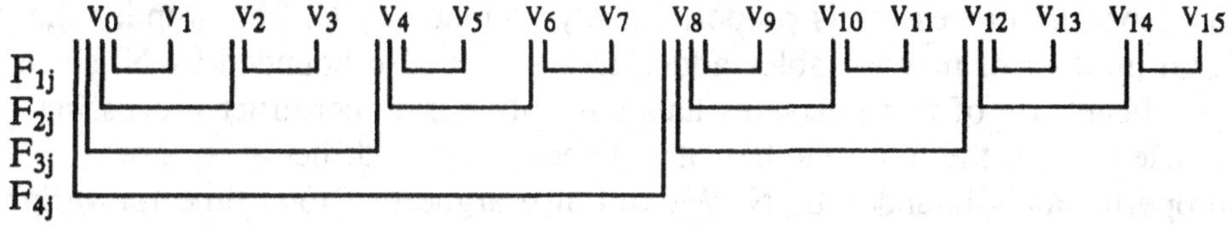

Assume merge processes the multiequations of $\mathcal{F}_n$ in row order. Then the computation ($\mathcal{E}$:= $\emptyset$; merge($\mathcal{E}$, $\mathcal{F}_n$)) causes the assignment find(w) := G in merge_1 to be executed $n \cdot 2^{n-1}$ times, so that time to execute the $m = 2^n - 1$ multiequations in $\mathcal{F}_n$ is bounded below by $m \cdot \log(m)$.

B. Assume each element of $\mathcal{F}$ contains exactly 2 variables. Let $\mathcal{E}' = \{E_1, ..., E_n\}$ be the result of the computing ($\mathcal{E}$:= $\emptyset$; merge($\mathcal{E}$, $\mathcal{F}$)). The time needed to compute $\mathcal{E}'$ is proportionately bounded by $k_1 \cdot \log(k_1) + ... + k_n \cdot \log(k_n)$, where k_i = size($E_i \cap V$), for each i.

C. The worst-case time complexity for ($\mathcal{E}$:= $\emptyset$; merge($\mathcal{E}$, $\mathcal{F}$)) is $O(m \cdot \log(m))$, where m = size(($\cup\mathcal{F}$) $\cap$ V).

Theorem 9.10. Assume the following:

- variable occurrence counts are well-defined in the instantiation system $\mathcal{A}$.
- simplify is deterministic.
- execution times for support routines have the following time complexities:
  t/v, $\sigma + \tau$, $t \in V$, $t \in C$, $u = v$, sub_both: 1;
  var(c), restricts, factors: wt(c); $c = d$, unifies_2c: wt(c) + wt(d);
  $t\sigma$: wt(t); $\sigma\tau$: wt(σ); $\sigma|U$: size(dom(σ)) + 1; $\rho^{\vee}$: size(dom(ρ)) + 1.

Then unify$_7$($\mathcal{D}$, δ) has an implementation that executes in time $N \cdot \log(N)$, where N = wt($\mathcal{D}$) + P, and where P is the total weight of all nonvariables returned by calls to restricts and unifies_2c.

Proof. As in previous proofs, we let E and $\mathcal{E}$ abbreviate the abstract images of J and $\mathcal{J}$, respectively. The two remaining implementation steps needed for the proof are,

a) defer the computation of δ, accumulating an indicated composition of the form $\delta = \theta_1 ... \theta_n$, and then perform the compositions from right to left.

b) implement restricts and unifies_2c in such a way that the new terms of the form var(cγ) added to $\mathcal{E}$ are such that var(cγ) = var_g(cγ); this may be done by calls to simplify, if necessary.

Execution time for the first main unify loop is linear in N; moreover, wt($\mathcal{E}$) is bounded by N on entry to the second loop. These facts will be clear from the analysis of the second main unify loop. We trivially generalize what is to be proved by assuming that unify$_7$ either terminates or is arbitrarily halted after q passes of the second main loop. In giving the proof, we associate calls to vc_reduce and cc_reduce with *construct processing* and associate calls to merge and vv_reduce with *variable processing*. We first show that the total time spent on construct processing is proportionately bounded by N. This implies that the total number of new variables added to $\mathcal{E} \cup \mathcal{F}$ is also bounded by N, since significant use of new variables happens only during construct processing. Consequently, the total number of multiequations and, hence, q, are proportionately bounded by N. We will also argue that total time for variable processing is bounded by $N \cdot \log(N)$. Finally, the proof will be completed by showing that total time break_cycle and the final computation of δ is also bounded by N.

To estimate the time for construct processing, we first introduce some notation. Extend the *construct weight* function, *cwt*, to substitutions and term systems by taking sums in the obvious way. Let $\mathcal{G}_{q+1}$ be the value of $\mathcal{S} \cup \mathcal{F}$ after the qth execution of the loop. For i = 1, ..., q, we introduce the following values pertaining to the ith execution of the body of the second main unify loop:

T_i — time spent on construct processing;
$\mathcal{S}_i$, $\mathcal{F}_i$ — the values of $\mathcal{S}$ and $\mathcal{F}$ at the top of loop;
E_i — the multiequation reduced;
P_i — weight of all nonvariables returned by calls to restricts and unifies_2c;
$\mathcal{G}_i = \mathcal{S}_i \cup \mathcal{F}_i$;
$\Delta_i = cwt(\mathcal{G}_i) - cwt(\mathcal{G}_{i+1}) + 2 \cdot P_i$.

Notice that

$$\Delta_1 + \dots + \Delta_q = cwt(\mathcal{G}_1) - cwt(\mathcal{G}_{q+1}) + 2 \cdot P \le cwt(\mathcal{G}_1) + 2 \cdot P < 2 \cdot N + 2 \cdot P \le 4 \cdot N,$$

using $\mathcal{G}_1 = \mathcal{S}_1$, $wt(\mathcal{S}_1) \le N$, and Proposition 6.9C. Hence, it suffices to show that T_i is proportionately bounded by Δ_i, for each i.

Our assumptions imply that a given call to vc_reduce takes time bounded by wt(c), where c is the construct produced by get_factor. Similarly, a call to cc_reduce is bounded by wt(c) + wt(d). If E_i is of the form $E_i = (\dots \approx v \approx c_1\alpha_1 \approx \dots \approx c_n\alpha_n)$, then T_i is bounded by $wt(c_1) + (wt(c_1) + wt(c_2)) + \dots + (wt(c_{n-1}) + wt(c_n)) < 2 \cdot (wt(c_1) + \dots + wt(c_n))$.

The net <u>decrease</u> in $cwt(\mathcal{G})$ which results from any given call to vc_reduce is of the form $cwt(t) - cwt(c\gamma) - cwt(\gamma|var(c)) - cwt(\alpha)$, which simplifies to $-2 \cdot cwt(\gamma|var(c))$, using Proposition 6.9C. Similarly, the net decrease resulting from a call to cc_reduce2 is

$$cwt(s) + cwt(t) - cwt(c\gamma) - cwt(\alpha) - cwt(\beta) - cwt(\gamma) = wt(c) - 2 \cdot cwt(\gamma).$$

If the unifiers produced during reduction of E_i are $\gamma_1, \dots, \gamma_n$, then the decrease resulting from deletion of the last term is $cwt(c_n\gamma_n)$, so that

$$\begin{aligned} &cwt(\mathcal{G}_i) - cwt(\mathcal{G}_{i+1}) \\ &\quad = -2 \cdot cwt(\gamma_1|var(c)) + \\ &\qquad\qquad (wt(c_1) - 2 \cdot cwt(\gamma_2)) + \dots + (wt(c_{n-1}) - 2 \cdot cwt(\gamma_n)) + cwt(c_n\gamma_n) \\ &\quad \ge (wt(c_1) + \dots + wt(c_n)) - 2 \cdot (cwt(\gamma_1|var(c)) + cwt(\gamma_2) + \dots + cwt(\gamma_n)) \\ &\quad \ge (wt(c_1) + \dots + wt(c_n)) - 2 \cdot P_i, \end{aligned}$$

so that $\Delta_i \ge (wt(c_1) + \dots wt(c_n))$. Hence, T_i is proportionately bounded by Δ_i, as required.

Let $\mathcal{H}$ be the multiset of all fragment multiequations produced during the computation. The total time spent in merge is proportional to the time needed to compute ($\mathcal{E} := \emptyset$; merge($\mathcal{E}$, $\mathcal{G}_1 \cup \mathcal{H}$)), because removal of base multiequations from $\mathcal{E}$ does not affect the merge computations. By Exercise 9.9C, this is bounded by $M \cdot \log(M)$, where M is the total number of variables in $(\mathcal{G}_1 \cup \mathcal{H}) \cap V$, and we know M is bounded by N.

Since simplify is deterministic, break_cycle will necessarily succeed in eliminating all spurious variables in var($\mathcal{E}$) the first time it is called. If it is called a second time, no new spurious variables will be found, since none are introduced. Consequently, no new base multiequations will be found, and break_cycle will raise cycle_failure. Hence, break_cycle is called at most twice. Since simplify is linear, total time spent in break_cycle is bounded by N.

We assumed that the construction of δ is deferred until the main loop terminates, and then evaluated as a right to left composition of the form $\delta = \theta_1(\theta_2(\ldots \theta_p))$, where the ith composition takes time proportional to $wt(\theta_i)$. Those θ_i produced by vv_reduce are variable-valued and their total weight is bounded by N, since q is. Those produced by vc_reduce are either variable-valued or contain nonvariables belonging to distinct $\mathcal{G}_i$, so their total weight is also bounded by N. Hence, the total weight of the θ_i is bounded by N. We conclude that time for the total computation is bounded by $N \cdot \log(N)$. □

SECTION 10
RELATED ISSUES NOT ADDRESSED

The following paragraphs point out what has not been achieved in previous sections, as a way of suggesting possible directions for future work.

To find unifiers in instantiation systems that have term occurrence cycles, it is necessary to unify equations of the form $v \approx t$, where $v \in var(t)$. Theorem 4.10 and the dimension concept in [SCHM88] suggest a reasonably general approach, one that is quite different from Huet's approach [HUET75].

In looking for systems with finite type intersections, we were led in Section 5 to ask which instantiation systems could be embedded in untyped systems, and when full embeddings and quotient mappings commute.

One might devise algorithms that use rational tree implementations for terms. A likely starting point for such an effort would be to find an analogue of Theorem 6.4 for rational trees. However, the same rational tree can easily implement two different terms.

Automatic theorem provers suggest at least two variants of the unification problem. The "matching" problem is often regarded as having more practical importance than the unification problem, because of its relation to backchaining. Forward chaining invites one to solve a problem slightly more general than unification, owing to the fact that any renaming of a theorem is itself a theorem: from r, $s \to t$, and $r\eta\sigma = s\sigma$, one may infer $t\sigma$, where η is a renaming. This observation suggests that there may be an advantage to combining renaming and unification in a single algorithm [cf LUSK82, APPL85].

The algorithm derived in Section 8 does not address the implementation of the simplify procedure. Developing rewrite-rule theory in the context of an arbitrary instantiation system could be highly instructive.

To actually use the unify_6 algorithm itself, we need an appropriate backtracking algorithm to handle control nondeterminacy. Ideally, one should be able to call the algorithm several times to enumerate a complete set of execution paths before failing.

Finally, there is a subtle unfairness regarding computational complexity results for unification algorithms. We assume that inputs have tree-like representations. But timing estimates for composing the unifier from its fragments rely implicitly on the use of shared subterms. No one has addressed, even for first-order term unification, the problem of determining computational complexity under the very reasonable assumption that both inputs and outputs have acyclic-graph implementations.

APPENDIX A
THE COMPILED UNIFICATION$_7$ SPECIFICATION

```
Generic package unification_2
    type O, V, S, C;
    with tσ, στ, σ + τ, σ|U, dom(σ), η^v, t/v, ∊, var(c),
          t ∈ V, t ∈ C, u = v, c = d, T_u = T_v, T_u ⊆ T_v;
          -- operations are given relative to the instantiation
          -- system (O, V, S, *) with construct basis C;
        procedure sub_both(T: out; T_1, T_2: in);
          -- exit T ⊆ T_1 ∩ T_2;
          -- feasibly if t ∈ T_1 ∩ T_2, then t ∈ T;
          -- time 1;
        procedure restricts(σ: out; c: in; T: in);
          -- exit cσ ∈ T and dom(σ) ⊆ var(c);
          -- feasibly if cτ ∈ T, then σ ≤_g≤_wt τ [var(c)];
          -- time wt(c);
        procedure unifies_2c(σ: out; c, d: in);
          -- entry var(c) ∩ var(d) = ∅;
          -- exit σ unifies_g c ≈ d and dom(σ) ⊆ var(c) ∪ var(d);
          -- feasibly if τ unifies_g c ≈ d, then σ ≤_g≤_wt τ [var(c ≈ d)]
          -- time wt(c) + wt(d);
        procedure factors(c, σ: out; t: in);
          -- entry t ∉ V;
          -- exit t = cσ and dom(σ) ⊆ var(c);
          -- feasibly true;
          -- time wt(c);
        procedure simplify(t: in out);
          -- exit t' =_g t and var(t') ⊆ var(t);
          -- feasibly var(t') = var_g(t) and wt(t') ≤ wt(t)
          -- time wt(t);
    end with;
is
    procedure unify(δ: out; 𝔇: in);
        -- exit δ unifies_g 𝔇;
        -- feasibly if τ unifies_g 𝔇, then σ ≤_g≤_wt τ [var(𝔇)]
        -- time N·log(N), where N = wt(𝔇) + P and P is the weight of all
        --      nonvariables returned by calls to restricts and unifies_2c;†
    clash_failure, cycle_failure: exception;
end unification_2;
```

† Additional caveats and restrictions apply regarding execution times.

```
Package body unification7 is

    type multieq_ptr = access
        record
            meq: multiequation;
            vdc: natural;
        end record;

    type container = array (variable) of multieq_ptr;

    type internal_system = multiset of multieq_ptr;


    procedure unify(δ: out := ε; 𝔇: in)
    is  𝔉 := 𝔇; E; J: multieq_ptr; U;
        find: container := (others => null);
        𝔍, base: internal_system := ∅;
    begin
        pick any U with U ⊇ var(𝔇);
        loop
            if possible, pick any E ∈ 𝔉 with E ∩ V = ∅;
            else exit end if;
            loop exit when size(E) = 1; cc_reduce(E, 𝔉, U); end loop;
            𝔉 := 𝔉 \ {E};
        end loop;
        init_vocs(find, 𝔍);
        loop
            if base = ∅ then
                merge(base, find, 𝔍, 𝔉);
                exit when 𝔍 = ∅;
                if base = ∅ then break_cycle(base, find, 𝔍); end if;
            end if;
            pick any J ∈ base;
            while size(J.meq ∩ V) > 1 loop vv_reduce(δ, J, U); end loop
            if J.meq ∩ V ≠ ∅ and J.meq \ V ≠ ∅ then
                vc_reduce(find, δ, J, 𝔉, U); end if;
            while size(J.meq \ V) > 1 loop cc_reduce2(find, J, 𝔉, U); end loop;
            𝔍 := 𝔍 \ {J};  base := base \ {J};
        end loop;
    end unify;
```

```
procedure vv_reduce(δ, E, U: in out)
is u, v, w; γ;
begin
    pick u ∈ E; pick v ∈ E \ {u};
    sub_both(T, T_u, T_v);
    pick any w ∈ {v, u} ∪ (V \ U) so that T_w = T; U := U ∪ {w};
    γ := w/u + w/v; δ := δγ;
    E := {w} ∪ (E \ {u, v});
end vv_reduce;

procedure cc_reduce(E, 𝔖, U: in out)
is s, t; c, d; α, β, γ;
begin
    pick s ∈ E \ V; pick t ∈ E \ (V ∪ {s});
    get_factor(c, α, s, ∅, U); get_factor(d, β, t, var(c), U);
    unifies_2c(γ, c, d); rename_cdm(γ, var(c ≈ d), U);
    𝔖 := 𝔖 \ {E}; E := {cγ} ∪ (E \ {s, t});
    𝔖 := 𝔖 ∪ {E} ∪ (γ ≈ α + β);
end cc_reduce;

procedure cc_reduce2(find, E, 𝔉, U: in out)
is s, t; c, d; α, β, γ;
begin
    pick s ∈ E \ V; pick t ∈ E \ (V ∪ {s});
    get_factor(c, α, s, ∅, U); get_factor(d, β, t, var(c), U);
    unifies_2c(γ, c, d); rename_cdm(γ, var(c ≈ d), U);
    E := {cγ} ∪ (E \ {s, t});
    if size(E) > 1 then add_vocs(find, 𝔍, cγ); end if;
    for v ∈ var(c ≈ d) loop
        add_vocs(find, 𝔍, vγ);
        𝔉 := 𝔉 ∪ {vγ ≈ v(α + β)};
    end loop;
end cc_reduce;
```

```
procedure vc_reduce(find, δ, E, 𝓕, U: in out)
is v; t; c; α, γ, ψ;
begin
    pick v ∈ E; pick t ∈ E \ V;
    get_factor(c, α, t, ∅, U);
    restricts(ψ, c, T_v);  rename_cdm(ψ, var(c), U);
    γ := cψ/v + ψ;
    δ := δ(γ|{v});
    E := {cγ} ∪ (E \ {v, t});
    if size(E) > 1 then add_vocs(find, 𝓘, cγ); end if;
    for v ∈ var(c) loop
        add_vocs(find, 𝓘, vγ);
        𝓕 := 𝓕 ∪ {vγ ≈ vα};
    end loop;
end vc_reduce;

procedure get_factor(c, α: out; t, W: in; U: in out)
is c; σ; η; μ;
begin
    factors(d, σ, t);
    pick any renaming μ such that
        dom(μ) ⊇ dom(σ) ∪ (W ∩ var(d)) and cdm(μ) ∩ U = ∅;
    c := dμ; α := (μ^vσ)|var(d); U := U ∪ var(c);
end get_factor;

procedure rename_cdm(γ: in out; W: in; U: in out)
is η := ∅;
begin
    pick any renaming η such that
        dom(η) ⊇ cdm(γ) \ (W \ dom(γ)) and cdm(η) ∩ U = ∅;
    γ := (γη)|dom(γ); U := U ∪ cdm(η);
end rename_cdm;
```

```
procedure break_cycle(base, find; 𝒮)
is s; F: multiequation; G: multieq_ptr;
begin
    for G in 𝒮 loop
        F := G.meq;
        for t in F \ V loop
            s := t; simplify(s); G.meq := G.meq \ {t} ∪ {s};
        end loop;
        G.vdc := 0;
    end loop;
    init_vocs(find, 𝒮); init_base(base, find, 𝒮);
    if base = ∅ then raise cycle_failure; end if;
end break_cycle;


procedure merge(base, find, 𝒮, ℱ: in out)
is begin
    while ℱ ≠ ∅ loop
        pick any F ∈ ℱ; ℱ := ℱ \ {F}; merge_1(ℰ, F);
    end loop;
end merge;


procedure merge_1(base, find, 𝒮: in out; F: in)
is  W := F ∩ V; G: multieq_ptr := new multieq_ptr(F \ V, 0);
    H, H2: multieq_ptr;
begin
    for v in W loop
        H := find(v);
        if H ≠ G then
            𝒮 := 𝒮 \ {H};
            if size(G.meq ∩ V) < size(H.meq ∩ V) then
                H2 := G; G := H; H := H2; end if;
            G.meq := G.meq ∪ H.meq; G.vdc := G.vdc + H.vdc - 1;
            for w in H.meq ∩ V loop find(w) := G; end loop;
        else G.vdc := G.vdc -1; end if;
    end loop;
    𝒮 := 𝒮 ∪ {G};
    if G.vdc = 0 then base := base ∪ {G}; end if;
end merge_1;
```

```
Procedure init_base(base: out := ø; 𝒮: in out)
is begin
   for J in 𝒮 loop
      If J.meq ∩ V = ø then base := base ∪ {J};
      else pick any v ∈ J.meq;
         if find(v).vdc = 0 then base := base ∪ {J}; end if;
      end if;
   end loop;
end init_base;

Procedure init_vocs(find: in out; 𝒮: in out)
is 𝒮̄: const := 𝒮;
begin
   for J in 𝒮̄ loop
      for t in J.meq \ V loop add_vocs(find, 𝒮, t); end loop;
   end loop;
end init_vocs;

Procedure add_vocs(find, 𝒮: in out; t)
is c; α;
begin
   if t ∈ V then
      if find(t) = null then
         find(t) := new multieq_ptr({t}, 1);
         𝒮 := 𝒮 ∪ {find(t)};
      else find(t).vdc := find(t).vdc + 1; end if;
   else
      factors(c, α, t);
      for v in var(c) loop add_vocs(find, 𝒮, vα); end loop;
   end if;
end add_vocs;

end unification₇;
```

REFERENCES

AHO83 Aho, A. V., Hopcroft, J. E., and Ullman, J. D., *Data Structures and Algorithms*, Addison-Wesley, 1983.

ANDR86 Andrews, P. B., *An Introduction to Mathematical Logic and Type Theory: To Truth Through Proof*, Academic Press, 1986.

APPL85 Applebaum, C. H., "Logic Machine Architecture Database Support System: Layer 1 User's Manual," MTP-256, The MITRE Corporation, November 1985.

BAUE85 Bauer, F. L., *The Munich Project CIP, Vol. I: The Wide Spectrum Language CIP-L*, Lecture Notes in Computer Science, No. 183, Springer-Verlag, 1985.

BLOO77 Bloom, S. L., Ginali, S., and Rutledge, J. D., "Scalar and Vector Iteration," *Journal of Computer and System Sciences*, Vol. 14, 1977, pp. 251-256.

BRUI72 de Bruin, N. G., "Lambda-Calculus Notation With Nameless Dummies, A Tool for Automatic Formula Manipulation, With Application to the Church-Rosser Theorem," *Indag. Math.* 34, 5 (1972), pp. 381-392.

BURC87 Bürckert, H.-J., Herold, A., and Schmidt-Schausz, M., "On Equational Theories, Unification and Decidability," *Proceedings of RTA'87*, Lecture Notes in Computer Science, No. 256, Springer-Verlag, 1987.

CLIF67 Clifford, A. H., and Preston, G. B., *The Algebraic Theory of Semigroups, Vol. II*, Mathematical Surveys No. 7, American Mathematical Society, 1967.

CLOC81 Clocksin, W. F. and Mellish, C. S, *Programming in Prolog*, Springer-Verlag, 1981.

CORB83 Corbin, J., and Bidoit, M., "A Rehabilitation of Robinson's Unification Algorithm," *Information Processing 83*, Elsevier Science, 1983.

COUR83 Courcelle, B., "Fundamental Properties of Infinite Trees," *Theoretical Computer Science*, Vol. 25, North-Holland, 1983, pp. 95-169.

CRAI87 Craigen, D., "m-Eves," CP-875402-26, *10th National Computer Security Conference*, NBS & NCSC, October 1987.

DOD83 United States Department of Defense, *Reference Manual for the Ada Programming Language*, ANSI/MIL-STD-1815A, 1983.

ELGO75 Elgot, C., "Monadic Computation and Iterative Algebraic Theories," in *Logic Colloquium* 73, edited by H. E. Rose, North-Holland, 1975, pp. 175-230.

ENDE72 Enderton, H. B., *A Mathematical Introduction to Logic*, Academic Press, 1972.

FARM90 Farmer, W. M., J. D., Guttman, and F. J. Thayer, "IMPS: An Interactive Mathematical Proof System," *Lecture Notes in Artificial Intelligence*, No. 449, Springer-Verlag, July 1990, pp. 653-654.

FARM90b Farmer, W. M., "A Partial Functions Version of Church's Simple Theory of Types," *Journal of Symbolic Logic*, Vol. 55, No. 3, September 1990, pp. 1269-1291.

FRAN88 Franzen, M., and Henschen, L. J., "A New Approach to Universal Unification and Its Application to AC-Unification," *9th International Conference on Automated Deduction*, Lecture Notes in Computer Science, Vol. 310, 1988.

FRAN88b Franzen, M., *A Unification Algorithm for Simple Equational Theories*, Doctorial Thesis, Northwestern University, forthcoming.

GINA79 Ginali, S., "Regular Trees and the Free Iterative Theory," *Journal of Computer and System Sciences*, Vol. 18, 1979, pp 228-242.

GALL88 Gallier, J. H., and Snyder, W., "Complete Sets of Transformations for General E-Unification," Draft, University of Pennsylvania, 1988.

GOGU87 Goguen, J. A., and Meseguer, J., "Models and Equality for Logical Programming," *TAPSOFT'87, Vol. 2*, Lecture Notes in Computer Science, Vol. 250, 1987.

GOGU88 Goguen, J. A., "What is Unification? A Categorical View of Substitution, Equation and Solution," SRI-CSL-88-2R2, SRI International, August 1988.

GOLD81 Goldfarb, W. D., "The Undecidability of the Second-Order Unification Problem," *Theoretical Computer Science* 13, 1981, pp. 225-230.

GOOD88 Good, D. I., DiVito, B. L., and Smith, M. K., *Using the Gypsy Methodology*, Computational Logic Inc., Austin Texas, January 1988.

GUTG76 Guttag, J. V., Horowitz, E., and Musser, D. R., "Abstract Data Types and Software Validation," ISI/RR-76-48, ARPA Order No. 2223, Information Sciences Institute, August, 1976.

GUTM85 Guttman, J. D., "Background Theory & Formal Semantics for the PVS Internal Language," MTP 255, The MITRE Corporation, September 1985.

HOWI76 Howie, J. M., *An Introduction to Semigroup Theory*, Academic Press, 1976.

HUET75 Huet, G. P., "A Unification Algorithm for Typed λ-Calculus," *Theoretical Computer Science,* Vol. 1, No. 1, North-Holland, 1975, pp. 27-57.

HUET75b Huet, G. P., "Unification in Typed Lambda Calculus," *Calculus and Computer Science Theory*, Lecture Notes in Computer Science, No. 37, Springer-Verlag, pp. 192–212, 1975.

HULL80 Hullot, J–M., "Canonical Forms and Unification," *Proceedings of the Fifth Conference on Automated Deduction,* Lecture Notes in Computer Science, Vol. 87, Springer-Verlag, 1980.

JAFF84 Jaffar, J., "Efficient Unification Over Infinte Terms," *New Generation Computing,* 2, 1984.

JECH78 Jech, T., *Set Theory,* Academic Press, 1978.

JENS73 Jensen, D. C., and T. Pietrzykowski, "Mechanizing ω-Order Type Theory Through Unification," Report CS-73-16, Dept. of Applied Analysis and Computer Science, University of Waterloo, 1973.

KNIG89 Knight, K., "Unification: A Multidisciplinary Survey," *ACM Computing Surveys,* Vol. 21, No. 1, March 1989, pp. 93–124.

LANK77 Lankford, D. S., and Ballantyne, A. M., *Decision Procedures for Simple Equational Theories with Permutative Axioms: Complete Sets of Reductions,* Report ATP-37, University of Texas, Austin, 1977.

LAWV63 Lawvere, F. W., "Functional Semantics of Algebraic Theories," *Proceedings of the National Academy of Sciences,* Vol. 50, No. 5, 1963.

LIVE75 Livesey, M. "Termination and Decidability Results for String Unification," CSM–12, 1975.

LOVE78 Loveland, D. W., *Automated Theorem Proving: A Logical Basis,* North Holland, 1978.

LUSK80 Lusk, E., and R. Ovebeek, "Data Structures and Control Architecture for the Implementation of Theorem-Proving Programs," *Proceedings of the Fifth Conference on Automated Deduction,* Lecture Notes in Computer Science, Vol. 87, Springer-Verlag, 1980.

LUSK82 Lusk, E., W. McCune, and R. Overbeek, "Logic Machine Architecture: Kernel Functions," *Proceedings of the Sixth Conference on Automated Deduction,* Lecture Notes in Computer Science, Vol. 138, Springer-Verlag, 1982.

MART82 Martelli, A., and U. Montanari, "An Efficient Unification Algorithm," *ACM Transactions on Programming Languages and Systems,* Vol. 4, No. 2, 1982.

MART86 Martelli, A., Moiso, C., and Rossi, G. F., "An Algorithm for Unification in Equational Theories," *IEEE 1986 Symposium on Logic Programming, IEEE, 1986.*

MCNU89 McNulty, G. F., "An Equational Logic Sampler," *Rewriting Techniques and Applications, RTA-89,* Lecture Notes in Computer Science No. 355, Springer-Verlag, April 1989, pp. 234-262.

MESE87 Meseguer, J., Goguen, J. A. and Smolka, G., "Order-Sorted Unification," *Proceedings of the Colloquium on the Resolution of Equations in Algebraic Structures,* Microelectronics and Computer Technology Corporation, May 1987.

NELS80 Nelson, G., and Oppen, D., "Fast Decision Procedures Based on Congruence Closure." *Journal of the Association for Computing Machinery,* 1980.

NERO59 Nerode, A., "Composita, Equations, and Freely Generated Algebras," *Transactions of The American Mathematical Society,* April 1959.

PARE70 Pareigis, B., *Categories and Functors,* Academic Press, 1970.

PIET72 Pietrzykowski, T, and D. C. Jensen, "A Complete Mechanization of (ω)-Order Type Theory," *National Conference Proceedings, Vol. 1,* ACM, 1972, pp. 82-92.

ROBI79 Robinson, J. A., *Logic, Form and Function: the Mechanization of Deductive Reasoning,* North-Holland, 1979.

RYDE86 Rydeheard, D. E., and R. M., Burstall, "A Categorical Unification Algorithm," *Proceedings of the Summer Conference on Category Theory and Computer Programming,* Lecture Notes in Computer Science, Vol. 240, Springer-Verlag, 1986.

RYDE87 Rydeheard, D. E., and J. G. Stell, "Foundations of Equational Deduction: A Categorical Treatment of Equational Proofs and Unification Algorithms," *Category Theory and Comoputer Science*, Lecture Notes in Computer Science, Vol. 283, Springer-Verlag, 1987, pp. 114–139.

SCHM86 Schmidt-Schauss, M., "Unification in Many-Sorted Equational Theories," *Proceedings of the 8th International Conference on Automated Deduction*, Lecture Notes in Computer Science, Vol. 230, Springer-Verlag, 1986.

SCHM88 Schmidt-Schauss, M., and J. H. Siekmann, "Unification Algebras: An Axiomatic Approach to Unification, Equation Solving and Constraint Solving," SEKI Report SR-88-23, University of Kaiserslautern, 1988.

SIEK75 Siekmann, J. H., "String Unification," University of Essex, CSM-7, 1975.

SIEK89 ________, "Unification Theory," *Journal of Symbolic Computation*, Academic Press, Vol. 7, 1989, pp. 207–274.

STEIN81 Steinbrugen, R., "The Composition of Schemes for Local Program Transformation," *Mathematical Models in Computer Systems: Proceedings of the Third Hungarian Computer Sciences Conference*, Akademiai Kiado, Budapest, 1981.

STICK81 Stickel, M. E., "A Unification Algorithm for Associative-Commutative Functions," *Journal of the Association for Computing Machinery*, Vol. 28, 1981.

STIL73 Stillman, R. B., "The concept of weak substitution in Theorem Proving," *Journal of the ACM*, Vol. 20, No. 4, ACM, 1973, pp. 648–667.

SZAB82 Szabö, P, *Unificationstheorie Erster Ordnung*, Dissertation, Universität Karlsruhe, 1982.

TARJ75 Tarjan, R. E., "On the Efficiency of a Good but Not Linear Set Merging Algorithm," *Journal of the ACM*, 22(2), 1975.

VANV75 van Vaalen, J. "An Extension of Unification to Substitutions with an Application to Automatic Theorem Proving," *Proceedings, Fourth International Joint Conference on Artificial Intellegence*, 1975, pp. 77-82.

WALT86 Walther, Ch., "A Classification of Many-Sorted Unification Algorithms," *Proceedings of the 8th International Conference on Automated Deduction*, Lecture Notes in Computer Science, Vol. 230, Springer-Verlag, 1986.

WILL87 Williams, J. G., "Formally Grounded Software Reasoning Systems: Theory and Requirements," MTP-266, The MITRE Corporation, April 1987.

WILL90 _______, "On the Formalization of Semantic Conventions," *Journal of Symbolic Logic*, Vol. 55, No. 1, March 1990, pp. 220–243.

Lecture Notes in Artificial Intelligence (LNAI)

Vol. 345: R.T. Nossum (Ed.), Advanced Topics in Artificial Intelligence. VII, 233 pages. 1988.

Vol. 346: M. Reinfrank, J. de Kleer, M.L. Ginsberg, E. Sandewall (Eds.), Non-Monotonic Reasoning. Proceedings, 1988. XIV, 237 pages. 1989.

Vol. 347: K. Morik (Ed.), Knowledge Representation and Organization in Machine Learning. XV, 319 pages. 1989.

Vol. 353: S. Hölldobler, Foundations of Equational Logic Programming. X, 250 pages. 1989.

Vol. 383: K. Furukawa, H. Tanaka, T. Fujisaki (Eds.), Logic Programming '88. Proceedings, 1988. IX, 251 pages. 1989.

Vol. 390: J.P. Martins, E.M. Morgado (Eds.), EPIA 89. Proceedings. 1989. XII, 400 pages. 1989.

Vol. 395: M. Schmidt-Schauß, Computational Aspects of an Order-Sorted Logic with Term Declarations. VIII, 171 pages. 1989.

Vol. 397: K.P. Jantke (Ed.), Analogical and Inductive Inference. Proceedings, 1989. IX, 338 pages. 1989.

Vol. 406: C.J. Barter, M.J. Brooks (Eds.), AI '88. Proceedings, 1988. VIII, 463 pages. 1990.

Vol. 418: K.H. Bläsius, U. Hedtstück, C.-R. Rollinger (Eds.), Sorts and Types in Artificial Intelligence. Proceedings, 1989. VIII, 307 pages. 1990.

Vol. 419: K. Weichselberger, S. Pöhlmann, A Methodology for Uncertainty in Knowledge-Based Systems. VIII, 132 pages. 1990.

Vol. 422: B. Nebel, Reasoning and Revision in Hybrid Representation Systems. XII, 270 pages. 1990.

Vol. 437: D. Kumar (Ed.), Current Trends in SNePS – Semantic Network Processing System. Proceedings, 1989. VII, 162 pages. 1990.

Vol. 444: S. Ramani, R. Chandrasekar, K.S.R. Anjaneyulu (Eds.), Knowledge Based Computer Systems. Proceedings, 1989. X, 546 pages. 1990.

Vol. 446: L. Plümer, Termination Proofs for Logic Programs. VIII, 142 pages. 1990.

Vol. 449: M.E. Stickel (Ed.), 10th International Conference on Automated Deduction. Proceedings, 1990. XVI, 688 pages. 1990.

Vol. 451: V. Mařík, O. Stepánková, Z. Zdráhal (Eds.), Artificial Intelligence in Higher Education. Proceedings, 1989. IX, 247 pages. 1990.

Vol. 459: R. Studer (Ed.), Natural Language and Logic. Proceedings, 1989. VII, 252 pages. 1990.

Vol. 462: G. Gottlob, W. Nejdl (Eds.), Expert Systems in Engineering. Proceedings, 1990. IX, 260 pages. 1990.

Vol. 465: A. Fuhrmann, M. Morreau (Eds.), The Logic of Theory Change. Proceedings, 1989. X, 334 pages. 1991.

Vol. 475: P. Schroeder-Heister (Ed.), Extensions of Logic Programming. Proceedings, 1989. VIII, 364 pages. 1991.

Vol. 476: M. Filgueiras, L. Damas, N. Moreira, A.P. Tomás (Eds.), Natural Language Processing. Proceedings, 1990. VII, 253 pages. 1991.

Vol. 478: J. van Eijck (Ed.), Logics in AI. Proceedings. 1990. IX, 562 pages. 1991.

Vol. 481: E. Lang, K.-U. Carstensen, G. Simmons, Modelling Spatial Knowledge on a Linguistic Basis. IX, 138 pages. 1991.

Vol. 482: Y. Kodratoff (Ed.), Machine Learning – EWSL-91. Proceedings, 1991. XI, 537 pages. 1991.

Vol. 513: N. M. Mattos, An Approach to Knowledge Base Management. IX, 247 pages. 1991.

Vol. 515: J. P. Martins, M. Reinfrank (Eds.), Truth Maintenance Systems. Proceedings, 1990. VII, 177 pages. 1991.

Vol. 517: K. Nökel, Temporally Distributed Symptoms in Technical Diagnosis. IX, 164 pages. 1991.

Vol. 518: J. G. Williams, Instantiation Theory. VIII, 133 pages. 1991.

Lecture Notes in Computer Science

Vol. 483: G. Rozenberg (Ed.), Advances In Petri Nets 1990. VI, 515 pages. 1991.

Vol. 484: R. H. Möhring (Ed.), Graph-Theoretic Concepts In Computer Science. Proceedings, 1990. IX, 360 pages. 1991.

Vol. 485: K. Furukawa, H. Tanaka, T. Fullsaki (Eds.), Logic Programming '89. Proceedings, 1989. IX, 183 pages. 1991. (Subseries LNAI).

Vol. 486: J. van Leeuwen, N. Santoro (Eds.), Distributed Algorithms. Proceedings, 1990. VI, 433 pages. 1991.

Vol. 487: A. Bode (Ed.), Distributed Memory Computing. Proceedings, 1991. XI, 506 pages. 1991.

Vol. 488: R. V. Book (Ed.), Rewriting Techniques and Applications. Proceedings, 1991. VII, 458 pages. 1991.

Vol. 489: J. W. de Bakker, W. P. de Roever, G. Rozenberg (Eds.), Foundations of Object-Oriented Languages. Proceedings, 1990. VIII, 442 pages. 1991.

Vol. 490: J. A. Bergstra, L. M. G. Feljs (Eds.), Algebraic Methods 11: Theory, Tools and Applicatlons. VI, 434 pages. 1991.

Vol. 491: A. Yonezawa, T. Ito (Eds.), Concurrency: Theory, Language, and Architecture. Proceedings, 1989. VIII, 339 pages. 1991.

Vol. 492: D. Sriram, R. Logcher, S. Fukuda (Eds.), Computer-Aided Cooperative Product Development. Proceedings, 1989 VII, 630 pages. 1991.

Vol. 493: S. Abramsky, T. S. E. Maibaum (Eds.), TAPSOFT '91. Volume 1. Proceedings, 1991. VIII, 455 pages. 1991.

Vol. 494: S. Abramsky, T. S. E. Maibaum (Eds.), TAPSOFT '91. Volume 2. Proceedings, 1991. VIII, 482 pages. 1991.

Vol. 495: 9. Thalheim, J. Demetrovics, H.-D. Gerhardt (Eds.), MFDBS '91. Proceedings, 1991. VI, 395 pages. 1991.

Vol. 496: H.-P. Schwefel, R. Männer (Eds.), Parallel Problem Solving from Nature. Proceedings, 1991. XI, 485 pages. 1991.

Vol. 497: F. Dehne, F. Fiala. W.W. Koczkodaj (Eds.), Advances in Computing and Intormation - ICCI '91 Proceedings, 1991. VIII, 745 pages. 1991.

Vol. 498: R. Andersen, J. A. Bubenko jr., A. Sølvberg (Eds.), Advanced Information Systems Engineering. Proceedings, 1991. VI, 579 pages. 1991.

Vol. 499: D. Christodoulakis (Ed.), Ada: The Choice for '92. Proceedings, 1991. VI, 411 pages. 1991.

Vol. 500: M. Held, On the Computational Geometry of Pocket Machining. XII, 179 pages. 1991.

Vol. 501: M. Bidoit, H.-J. Kreowski, P. Lescanne, F. Orejas, D. Sannella (Eds.), Algebraic System Specification and Development. VIII, 98 pages. 1991.

Vol. 502: J. Bārzdiņš, D. Bjørner (Eds.), Baltic Computer Science. X, 619 pages. 1991.

Vol. 503: P. America (Ed.), Parallel Database Systems. Proceedings, 1990. VIII, 433 pages. 1991.

Vol. 504: J. W. Schmidt, A. A. Stogny (Eds.), Next Generation Information System Technology. Proceedings, 1990. IX, 450 pages. 1991.

Vol. 505: E. H. L. Aarts, J. van Leeuwen, M. Rem (Eds.), PARLE '91. Parallel Architectures and Languages Europe, Volume I. Proceedings, 1991. XV, 423 pages. 1991.

Vol. 506: E. H. L. Aarts, J. van Leeuwen, M. Rem (Eds.), PARLE '91. Parallel Architectures and Languages Europe, Volume II. Proceedings, 1991. XV, 489 pages. 1991.

Vol. 507: N. A. Sherwani, E. de Doncker, J. A. Kapenga (Eds.), Computing in the 90's. Proceedings, 1989. XIII, 441 pages. 1991.

Vol. 508: S. Sakata (Ed.), Applied Algebra, Algebraic Algorithms and Error-Correcting Codes. Proceedings, 1990. IX, 390 pages. 1991.

Vol. 509: A. Endres, H. Weber (Eds.), Software Development Environments and CASE Technology. Proceedings, 1991. VIII, 286 pages. 1991.

Vol. 510: J. Leach Albert, B. Monien, M. Rodríguez (Eds.), Automata, Languages and Programming. Proceedings, 1991. XII, 763 pages. 1991.

Vol. 511: A. C. F. Colchester, D.J. Hawkes (Eds.), Information Processing in Medical Imaging. Proceedings, 1991. XI, 512 pages. 1991.

Vol. 512: P. America (Ed.), ECOOP '91. European Conference on Object-Oriented Programming. Proceedings, 1991. X, 396 pages. 1991.

Vol. 513: N. M. Mattos, An Approach to Knowledge Base Management. IX, 247 pages. 1991. (Subseries LNAI).

Vol. 514: G. Cohen, P. Charpin (Eds.), EUROCODE '90. Proceedings, 1990. XI, 392 pages. 1991.

Vol. 515: J. P. Martins, M. Reinfrank (Eds.), Truth Maintenance Systems. Proceedings, 1990. VII, 177 pages. 1991. (Subseries LNAI).

Vol. 516: S. Kaplan, M. Okada (Eds.), Conditional and Typed Rewriting Systems. Proceedings, 1990. IX, 461 pages. 1991.

Vol. 517: K. Nökel, Temporally Distributed Symptoms in Technical Diagnosis. IX, 164 pages. 1991. (Subseries LNAI).

Vol. 518: J. G. Williams, Instantiation Theory. VIII, 133 pages. 1991. (Subseries LNAI).